FUSION 360 EXERCISES

200 PRACTICE DRAWINGS

SACHIDANAND JHA

Dear Reader,

Thank you for choosing **AUTODESK FUSION 360 EXERCISES** book. This book is part of a family of premium-quality CADIN360 books, all of which are written by Outstanding author who combine practical experience with a gift for teaching.

CADIN360 was founded in 2016. More than 3 years later, we're still committed to producing consistently exceptional books. With each of our titles, we're working hard to set a new standard for the industry. From the paper we print on, to the authors we work with, our goal is to bring you the best books available.

I hope you see all that reflected in these pages. I'd be very interested to hear your comments and get your feedback on how we're doing. Feel free to let me know what you think about this or any other CADin360 book by sending me an email at contactus@cadin360.com.

If you think you've found a technical error in this book, please visit
https://cadin360.com/contact-us/.
Customer feedback is critical to our efforts at CADIN360.

Best regards,

Sachidanand Jha
Founder & CEO, CADIN360

AUTODESK FUSION 360 EXERCISES

Preface

AUTODESK FUSION 360 EXERCISES

- ❖   This book contain 200 CAD practice exercises and drawings.

- ❖   This book does not provide step by step tutorial to design 3D models.

- ❖   Get all the Original CAD files on cadin360.com

- ❖   S.I Units is used.

- ❖   Predominantly used Third Angle Projection.

- ❖   This book is for **AUTODESK FUSION 360** and Other Feature-Based Modeling Software such as Inventor, Catia, SolidWorks, NX, Solid Edge, AutoCAD, PTC Creo etc.

- ❖   It is intended to provide Drafters, Designers and Engineers with enough CAD exercises for practice on **FUSION 360** .

- ❖   It includes almost all types of exercises that are necessary to provide, clear, concise and systematic information required on industrial machine part drawings.

- ❖   Third Angle Projection is intentionally used to familiarize Drafters, Designers and Engineers in Third Angle Projection to meet the expectation of world wide Engineering drawing print.

- ❖   Clear and well drafted drawing help easy understanding of the design.

- ❖   This book is for Beginner, Intermediate and Advance CAD users.

- ❖   These exercises are from Basics to Advance level.

- ❖   Each exercises can be assigned and designed separately.

- ❖   No Exercise is a prerequisite for another. All dimensions are in mm.

- ❖   Note: Assume any missing dimensions.

AUTODESK FUSION 360 EXERCISES files

Get all the original Autodesk **Fusion 360** Exercises (**200 CAD files**) used in this book. Download all 200 3D models and practice drawings or exercises on **cadin360.com**

Why should i buy AUTODESK FUSION 360 EXERCISES files?

❖ **G**et all **200** exercises: to know the steps envolved in creating these models.

❖ **Y**ou can suppress and unsuppress features, to know step by step modeling process of these exercises.

❖ **Y**ou can modify all the models used in this book as per your skills.

❖ **Y**ou can redesign all these exercises from scratch to improve your cad skills.

❖ **Y**ou can do drafting on these models and enhance your Drafting skills.

❖ **Y**ou can modify these models for 3D Printing.

❖ Use these original files to learn, practice, or even improve them to build better designs.

Get all the Original Autodesk fusion 360 Exercises files on **cadin360.com**

For more information visit:- **https://cadin360.com/autodesk-fusion-360-exercises/**

3D

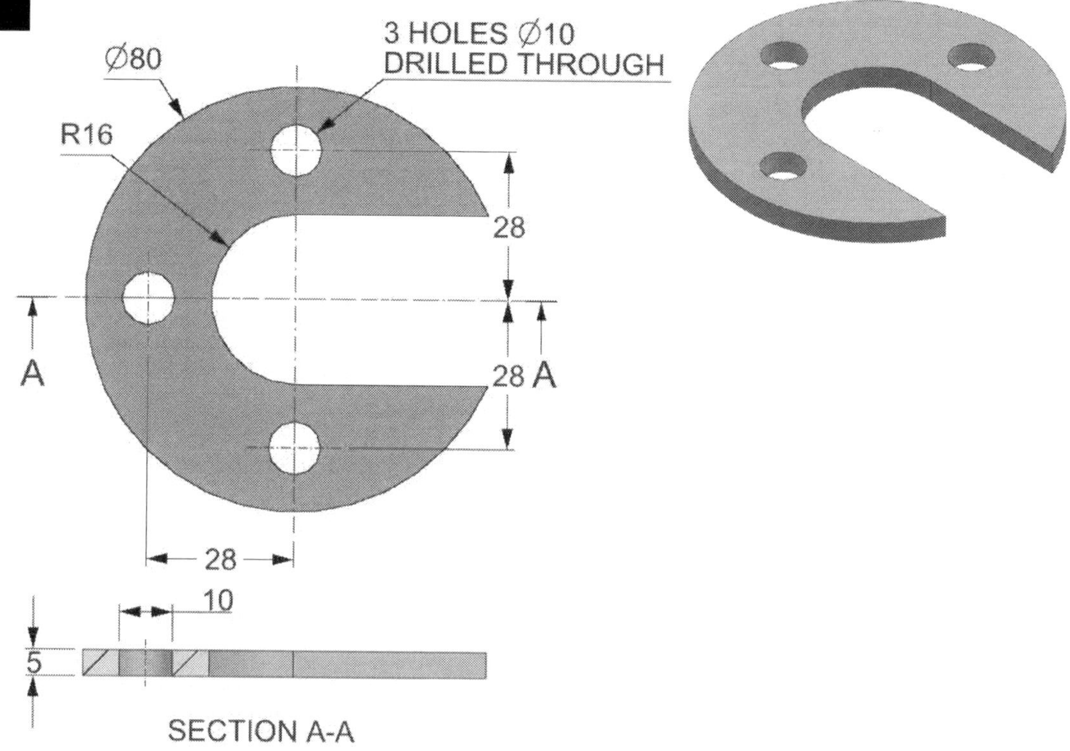
Ø80
R16
3 HOLES Ø10
DRILLED THROUGH
28
28
A
A
28
10
5
SECTION A-A

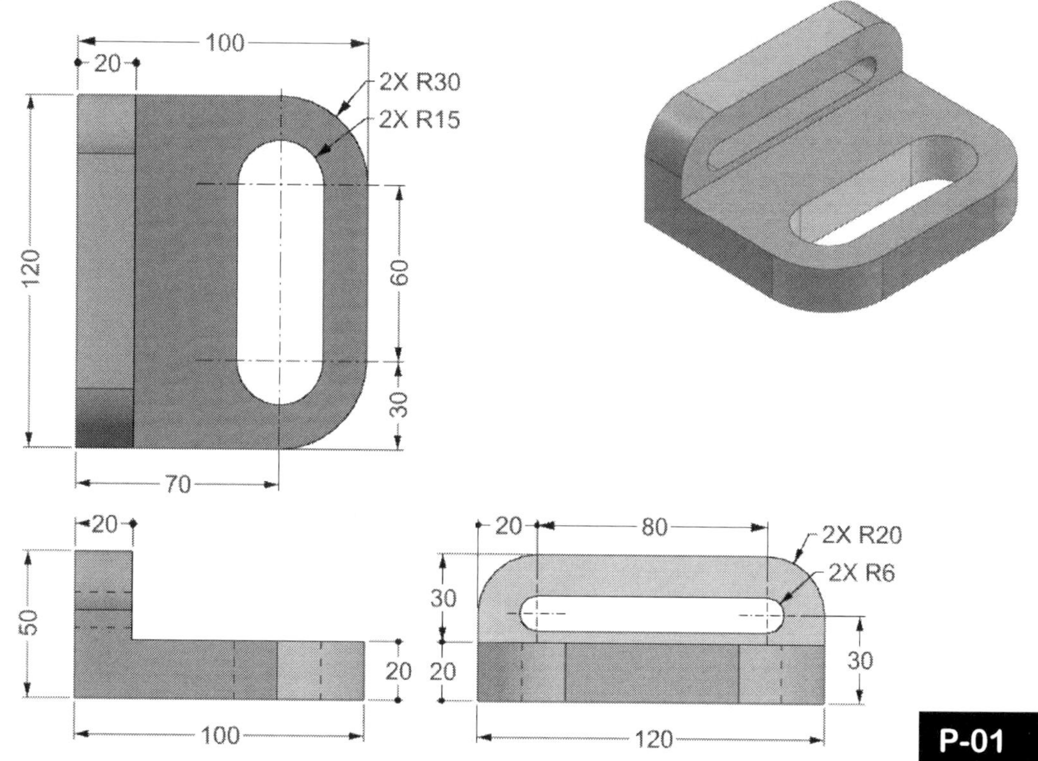
100
20
2X R30
2X R15
120
60
30
70
20
50
20
100
20
80
2X R20
2X R6
30
30
20
20
120

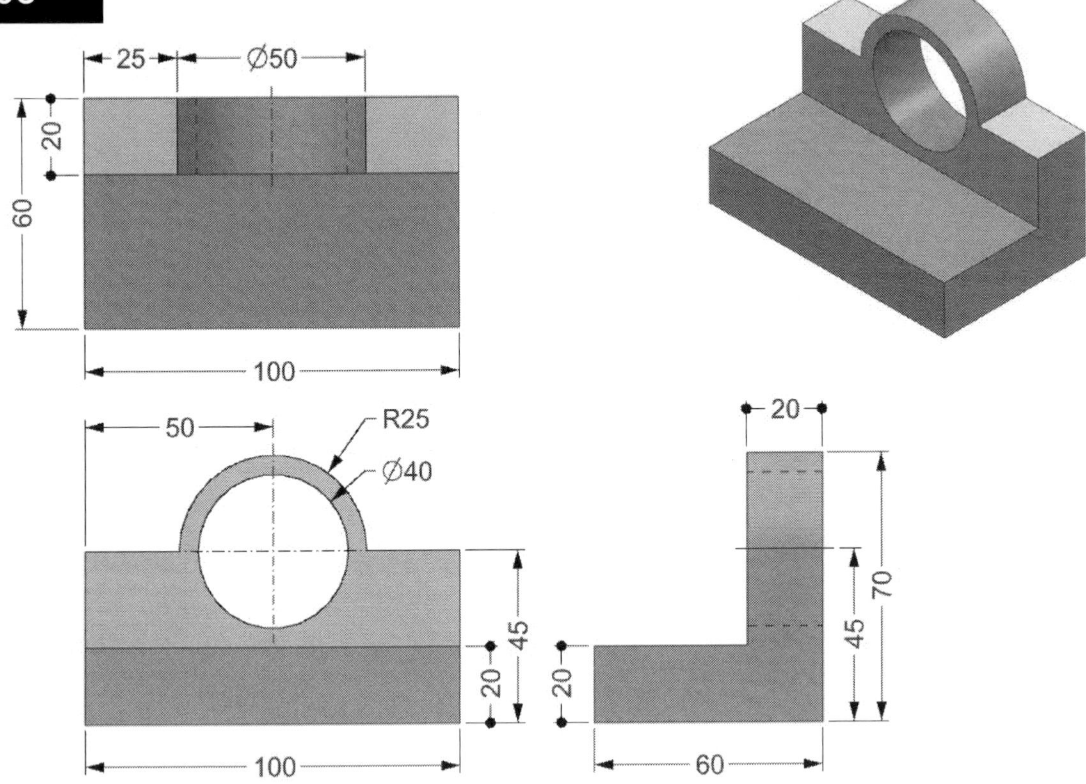

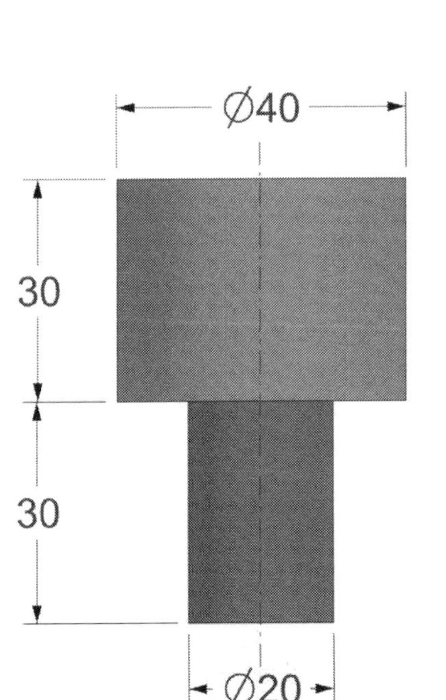

EX-05

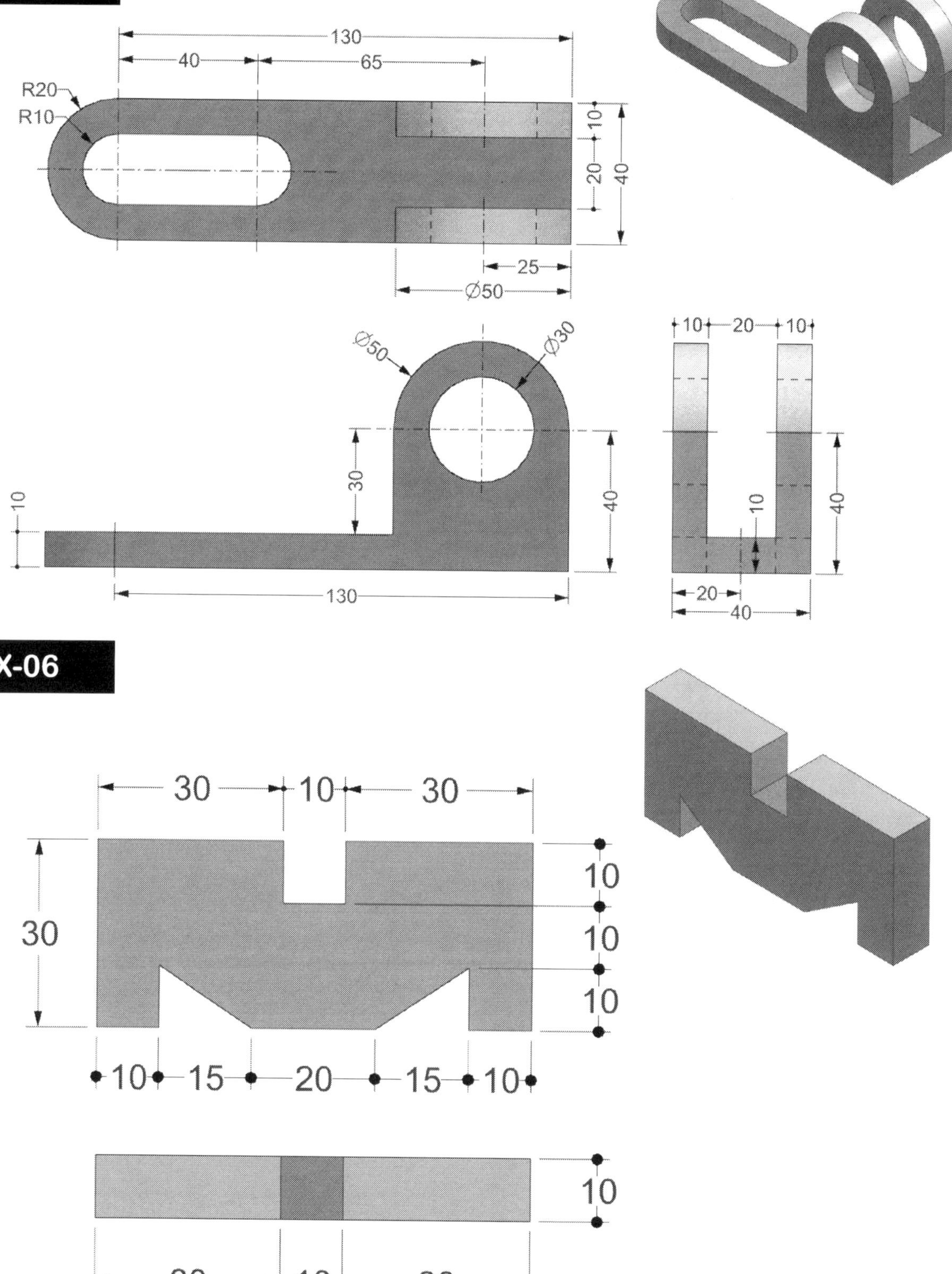
EX-05
130
40
65
R20
R10
10
20
40
25
Ø50
Ø50
Ø30
30
40
10
130
10
20
10
10
40
20
40
EX-06
30
10
30
30
10
10
10
10
15
20
15
10
10
30
10
30
P-03

EX-07

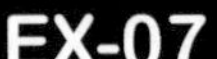

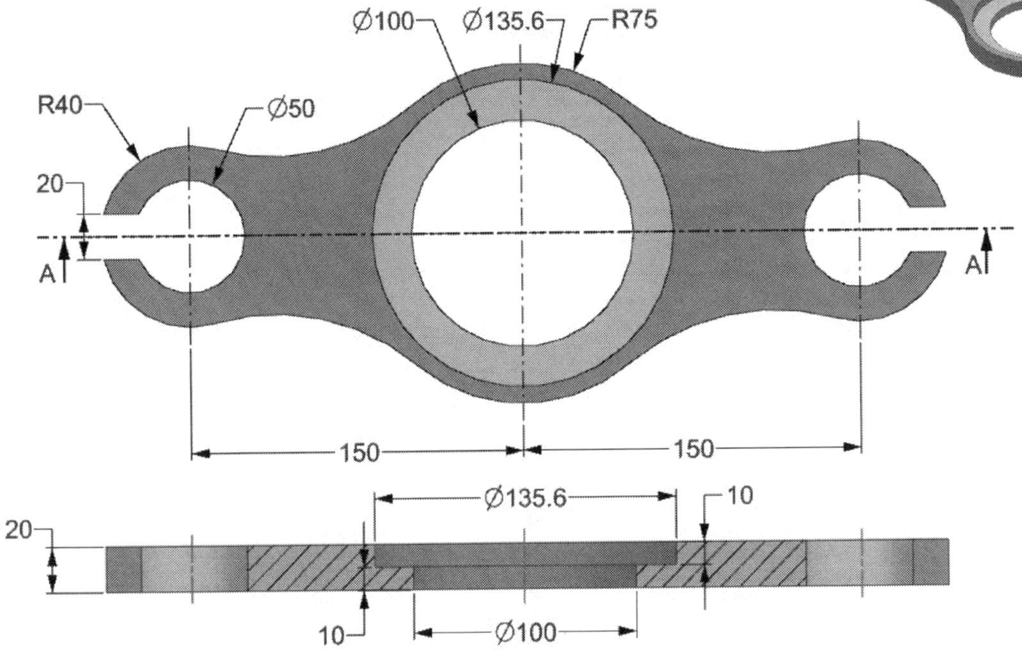

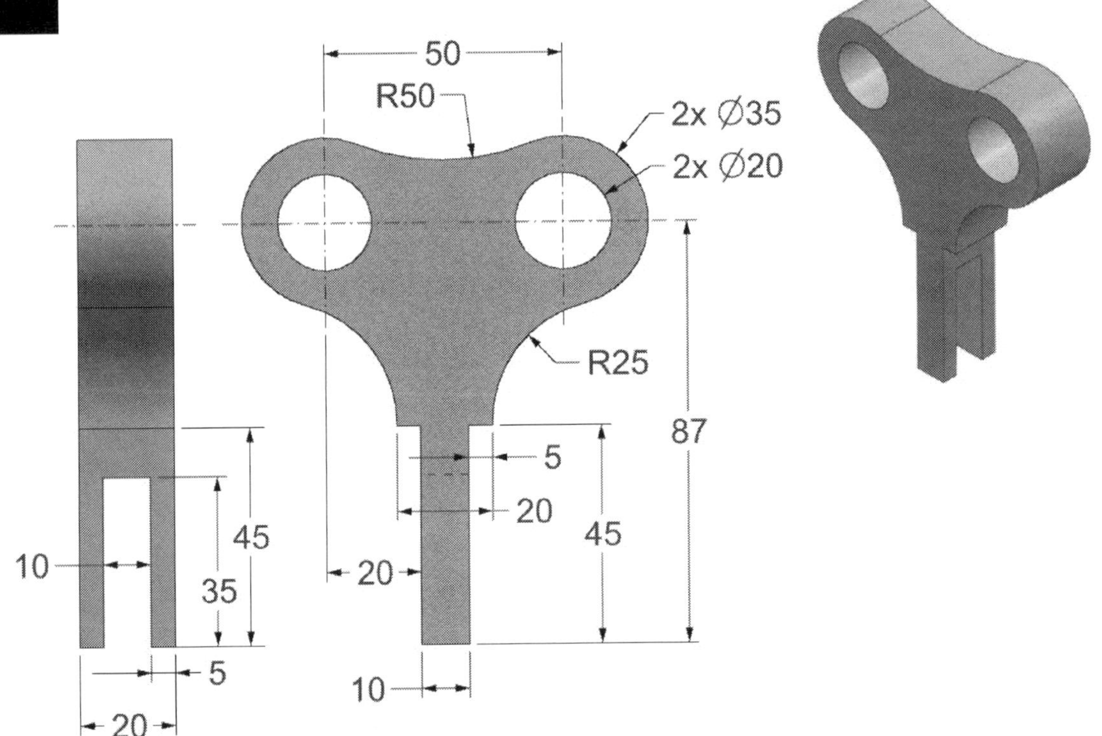

Ø100
Ø135.6
R75
R40
Ø50
20
A
A
150
150
Ø135.6
10
20
10
Ø100
SECTION A-A
(SCALE 1:1)
EX-08
50
R50
2x Ø35
2x Ø20
R25
87
5
20
45
10
45
35
10
5
20
20
10

P-04

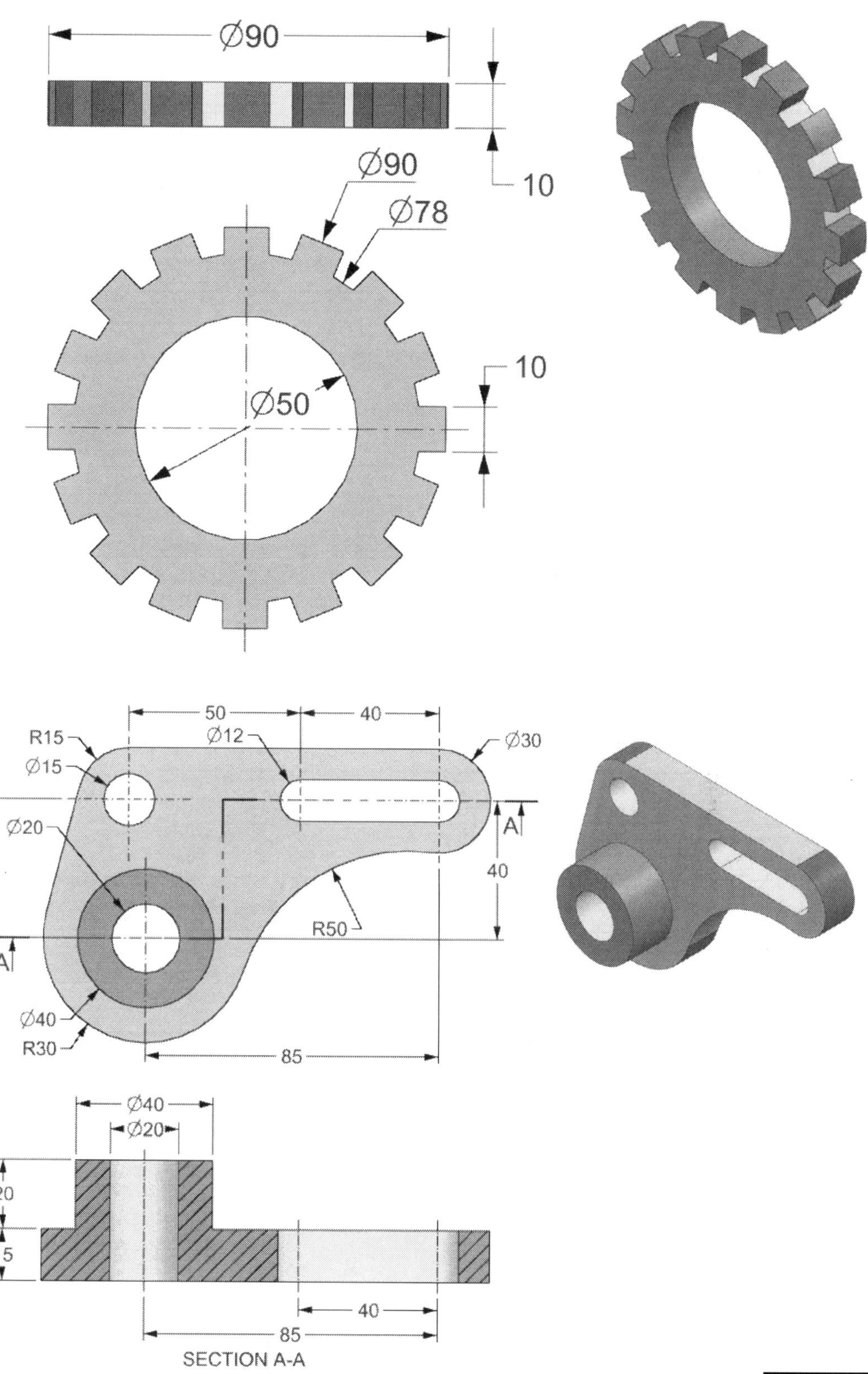

⌀90
10
⌀90
⌀78
⌀50
10
EX-10
50
40
R15
⌀12
⌀30
⌀15
⌀20
A
40
40
R50
A
R30
⌀40
85
⌀40
⌀20
20
15
40
85
SECTION A-A
(SCALE 1:1)

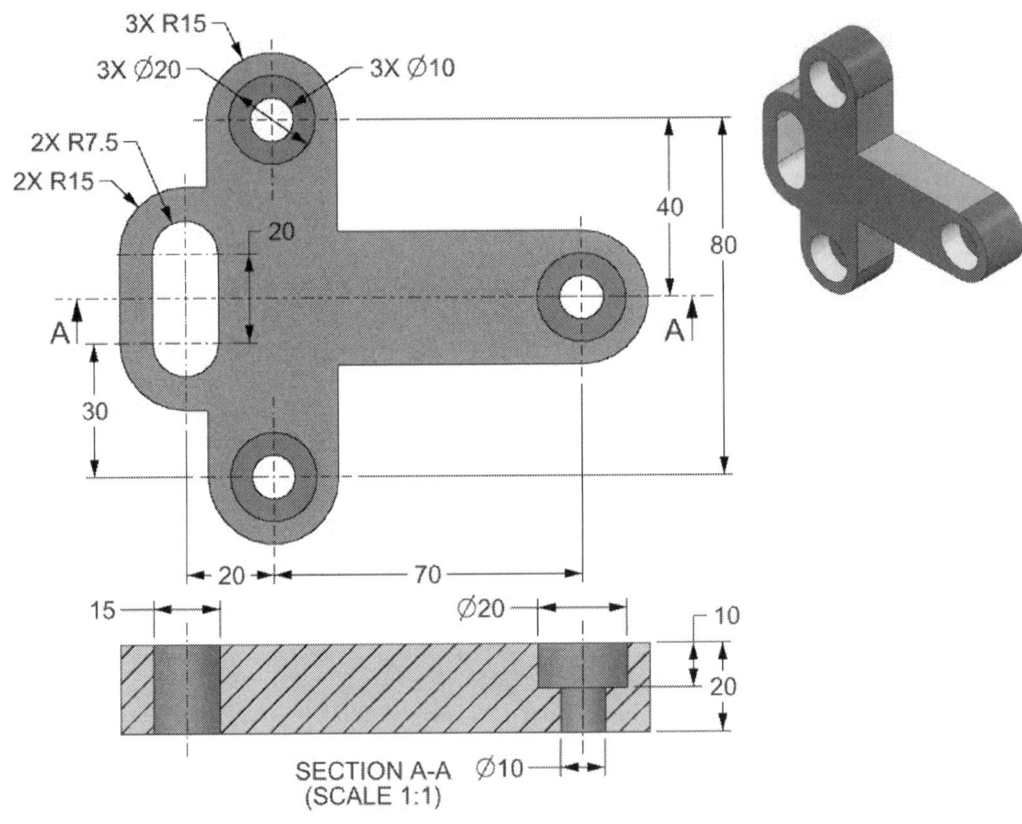

3X R15
3X Ø20
3X Ø10
2X R7.5
2X R15
20
40
80
30
20
70
15
Ø20
10
20
Ø10
SECTION A-A
(SCALE 1:1)
A
A

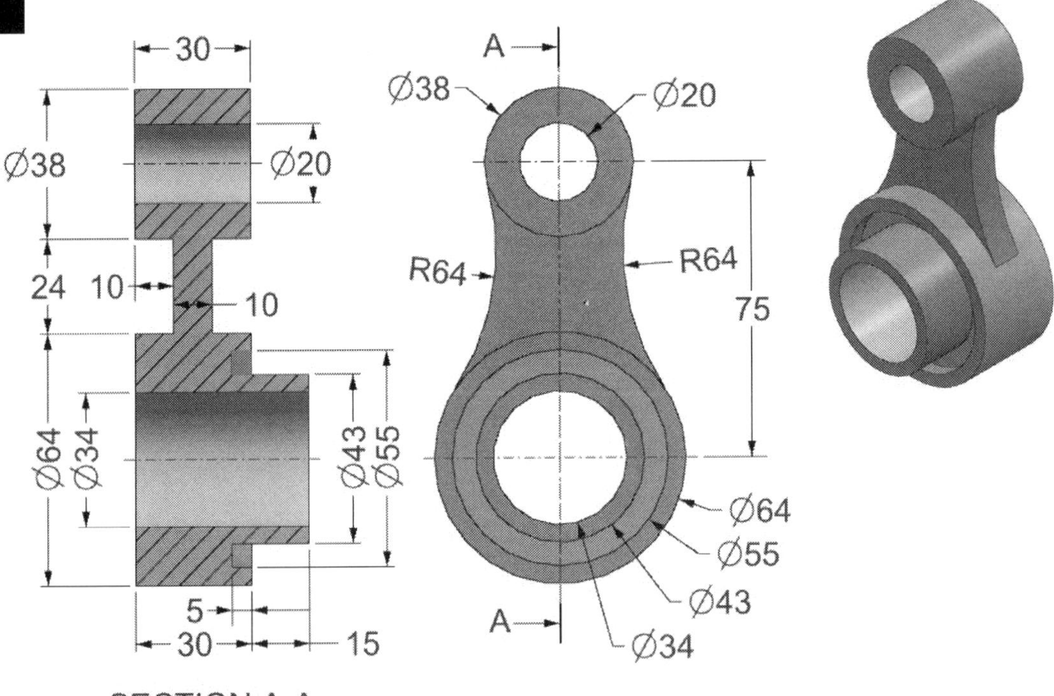

30
Ø38
Ø20
24
10
10
Ø64
Ø34
Ø43
Ø55
5
30
15
A
Ø38
Ø20
R64
R64
75
Ø64
Ø55
Ø43
Ø34
A
SECTION A-A
(SCALE 1:1)

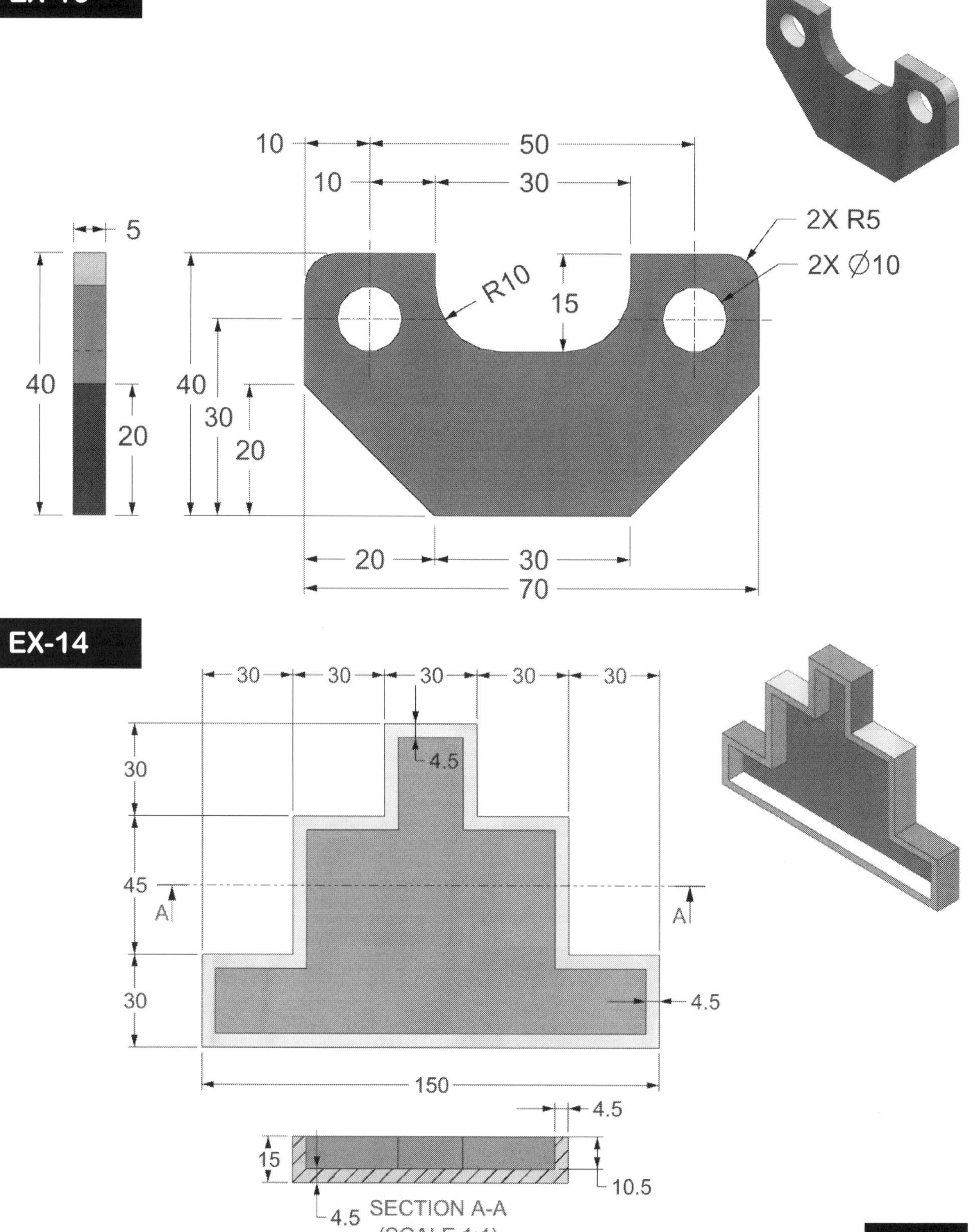

EX-13
5
40
20
10
50
10
30
2X R5
2X Ø10
R10
15
40
30
20
20
30
70
EX-14
30
30
30
30
30
30
4.5
30
45
A
A
30
4.5
150
4.5
15
10.5
4.5
SECTION A-A
(SCALE 1:1)
P-07

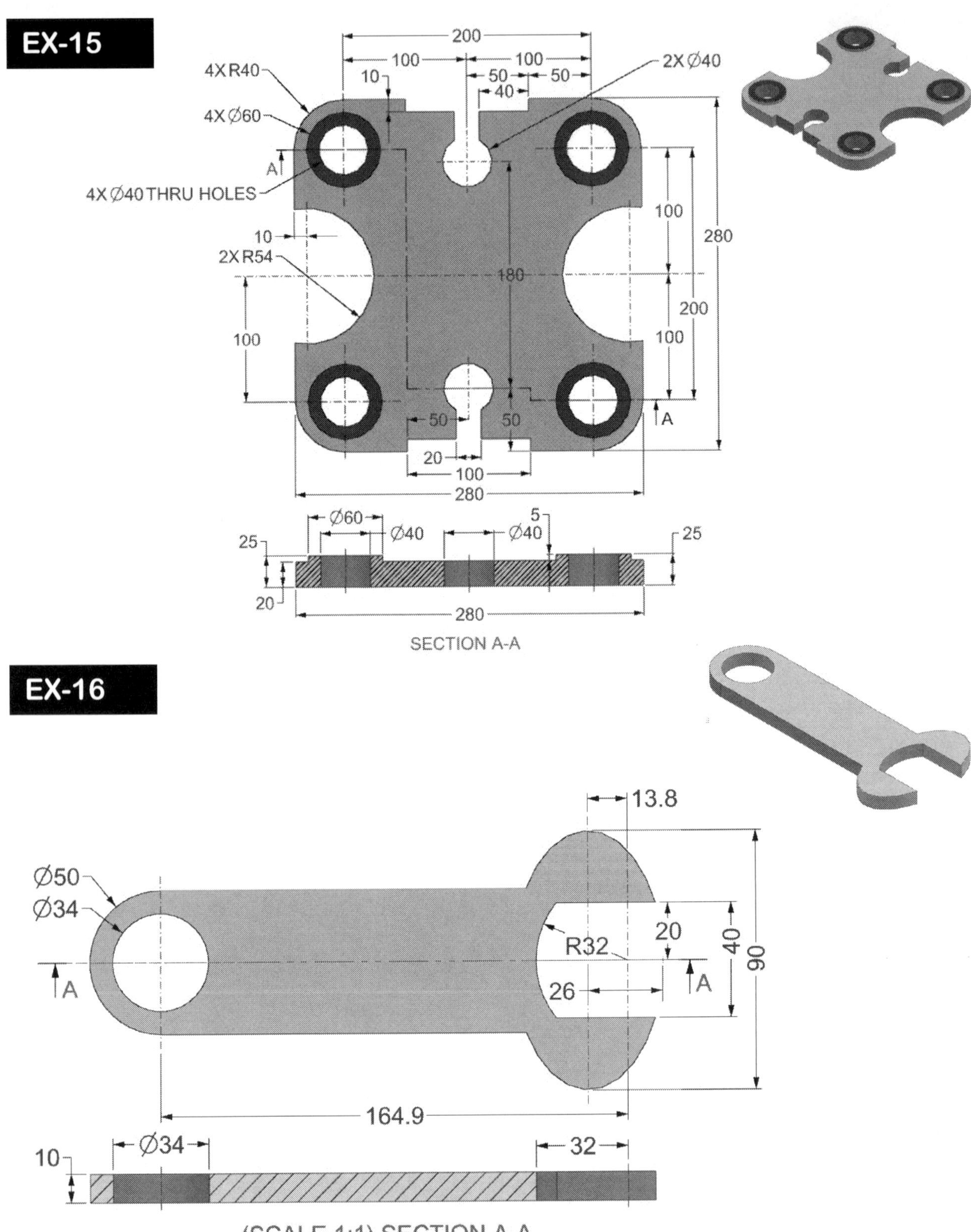

EX-15
200
100
100
4X R40
10
50
50
40
2X Ø40
4X Ø60
A
4X Ø40 THRU HOLES
10
100
280
2X R54
180
100
200
100
50
50
A
20
100
280
Ø60
Ø40
5
Ø40
25
25
20
280
SECTION A-A
EX-16
13.8
Ø50
Ø34
R32
20
40
90
26
A
A
164.9
Ø34
32
10
(SCALE 1:1) SECTION A-A
P-08

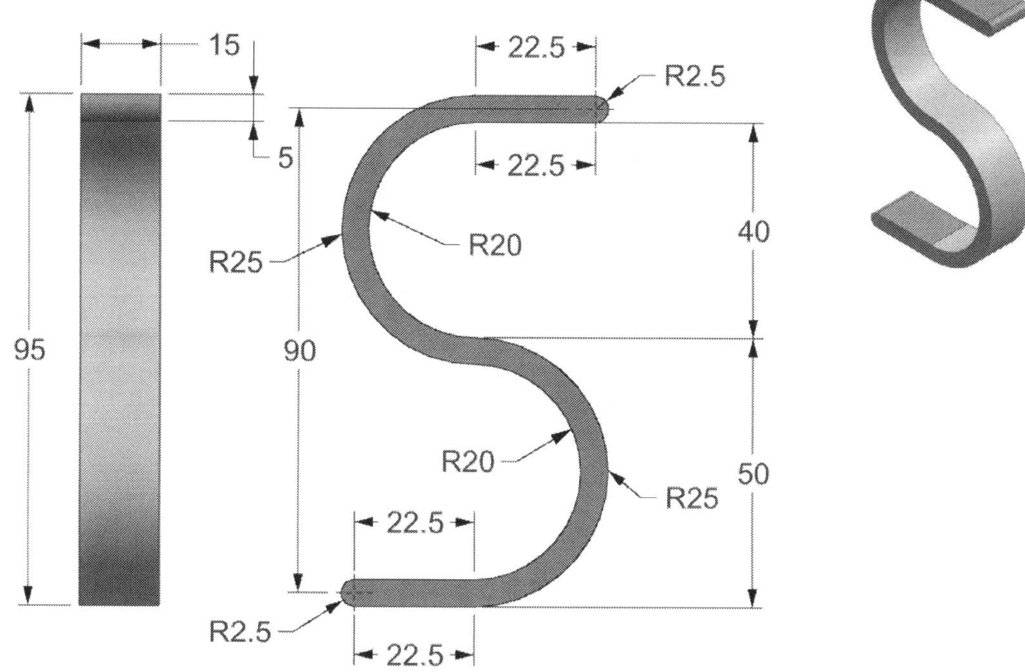

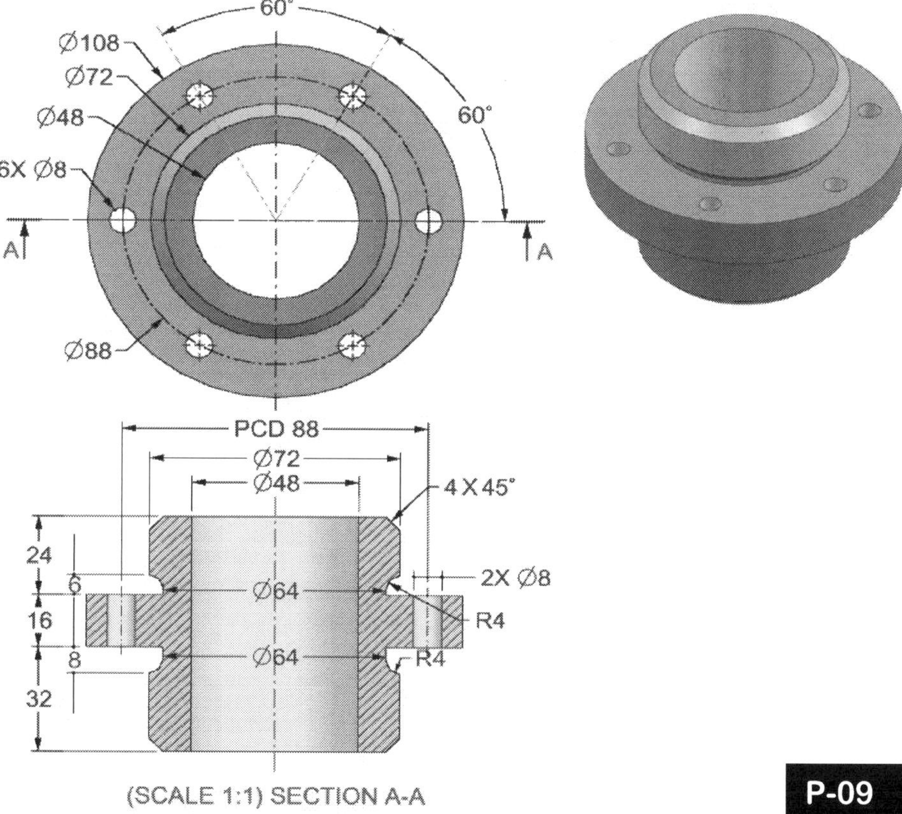

(SCALE 1:1) SECTION A-A

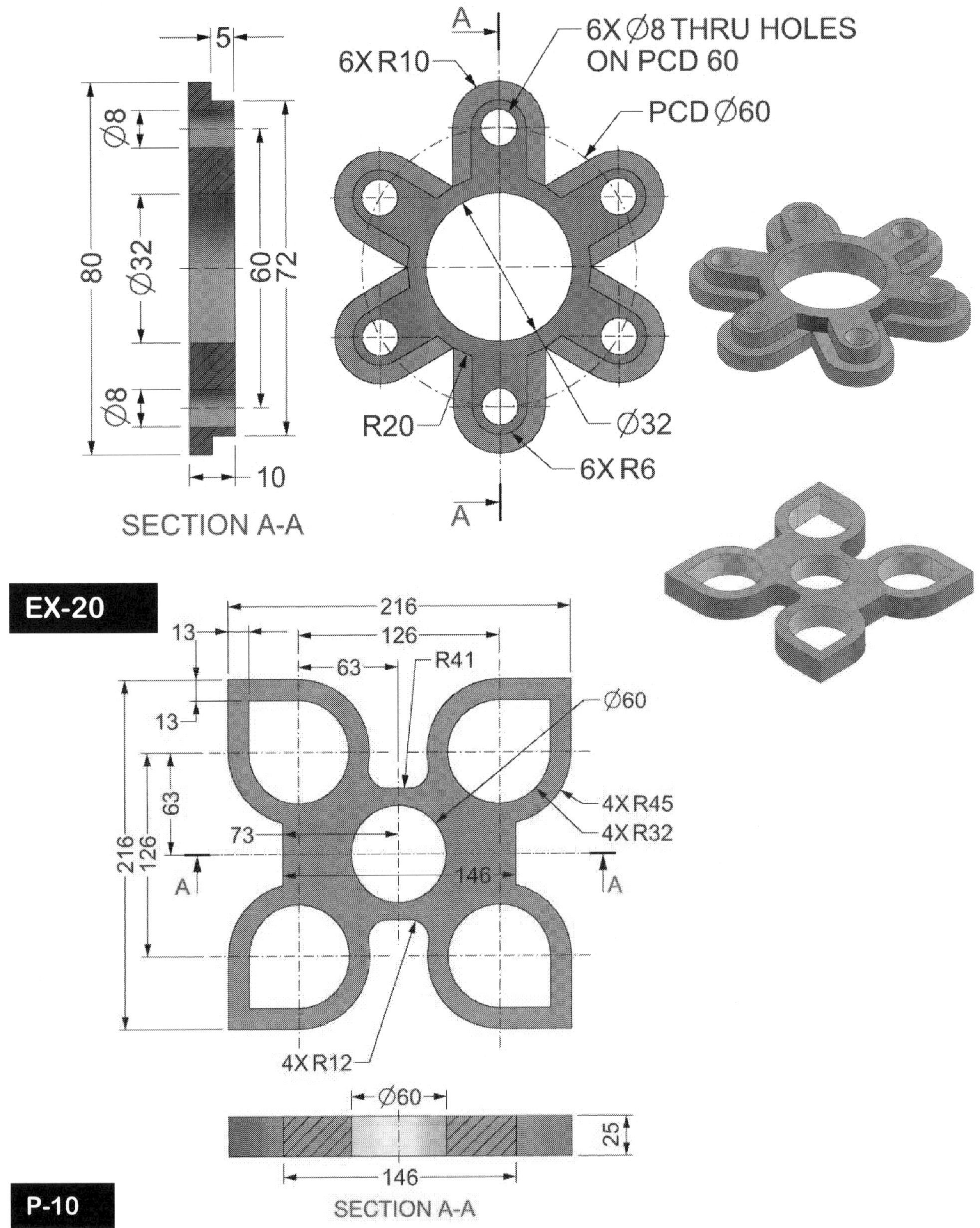
5
Ø8
Ø32
Ø8
80
60
72
10
SECTION A-A
A
6X R10
6X Ø8 THRU HOLES
ON PCD 60
PCD Ø60
R20
Ø32
6X R6
A
EX-20
216
13
126
63
R41
Ø60
13
63
216
126
73
4X R45
4X R32
146
A
A
4X R12
Ø60
25
146
SECTION A-A

EX-21 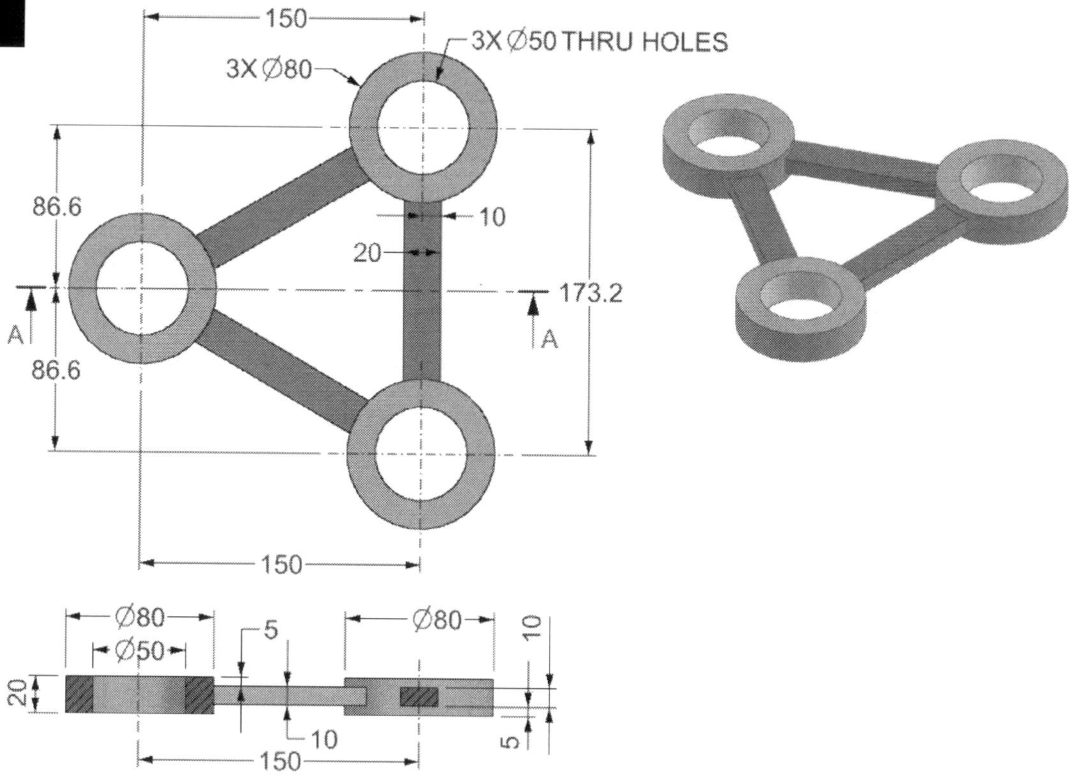
150
3X Ø50 THRU HOLES
3X Ø80
86.6
10
20
173.2
A
86.6
A
150
Ø80
Ø50
5
Ø80
10
20
10
5
150
SECTION A-A

EX-22 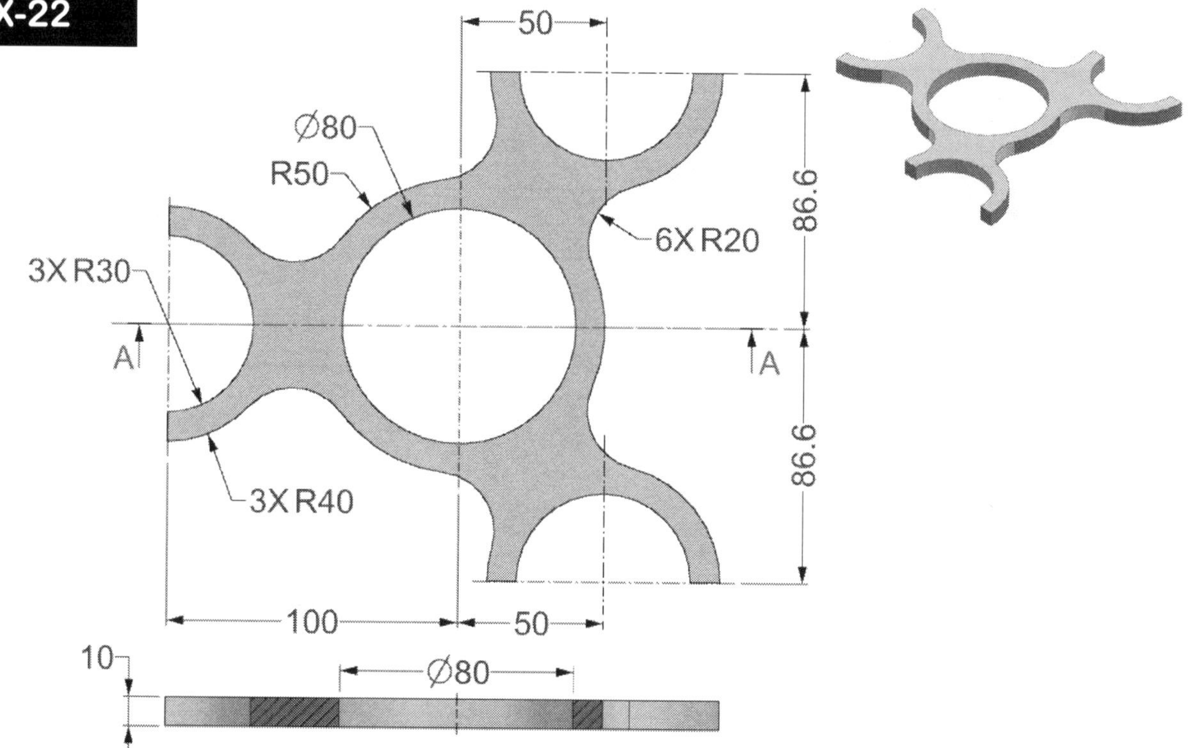
50
Ø80
R50
86.6
6X R20
3X R30
A
A
3X R40
86.6
100
50
10
Ø80
SECTION A-A

P-11

EX-23
2X R50
2X R35
214
15
15
214
214
214
20
214
EX-24
30
60
60
20
100
60
100
40
20
30
60
60
10
10
10
10
10
70
150
10
50
70
20
60
20
100
P-12

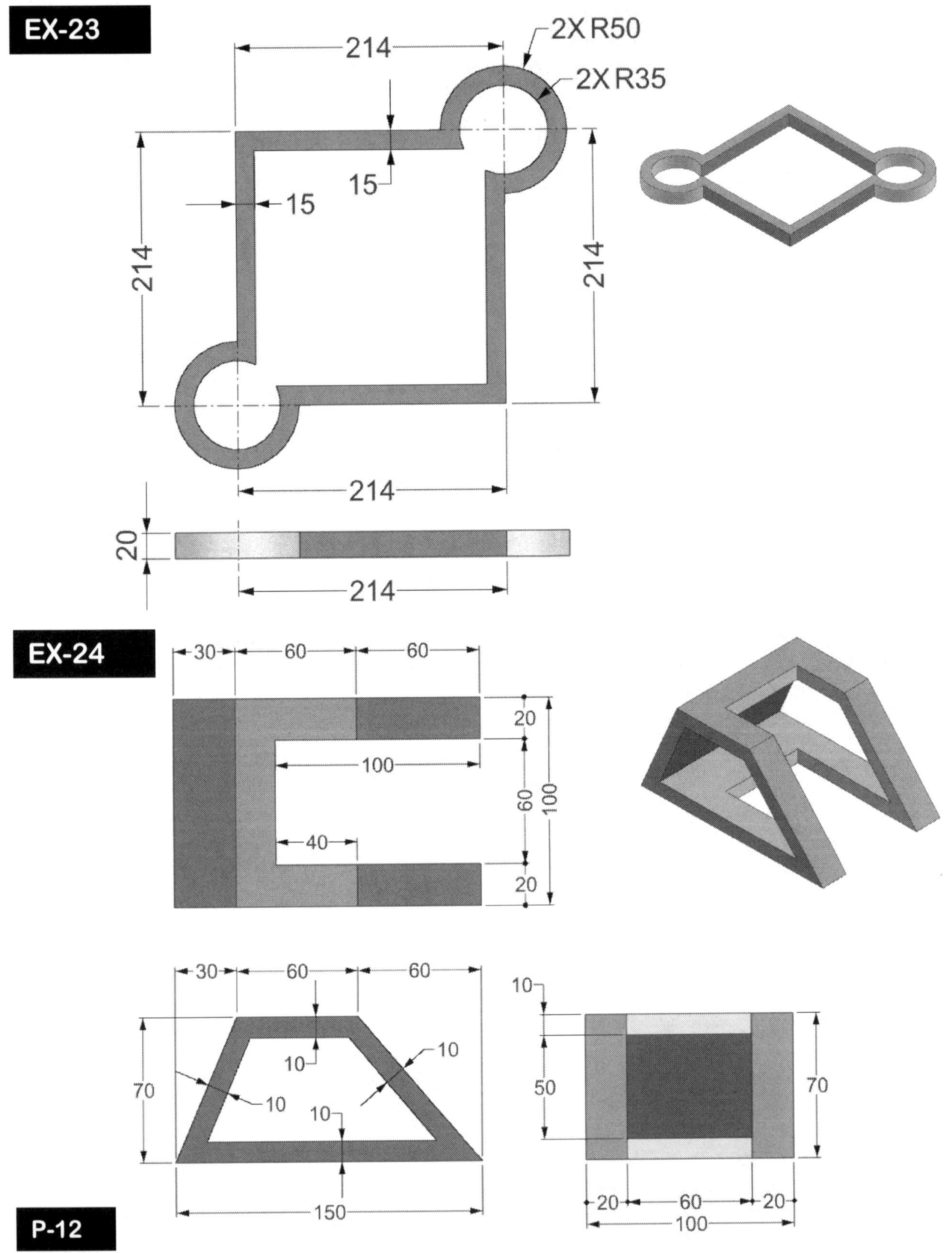

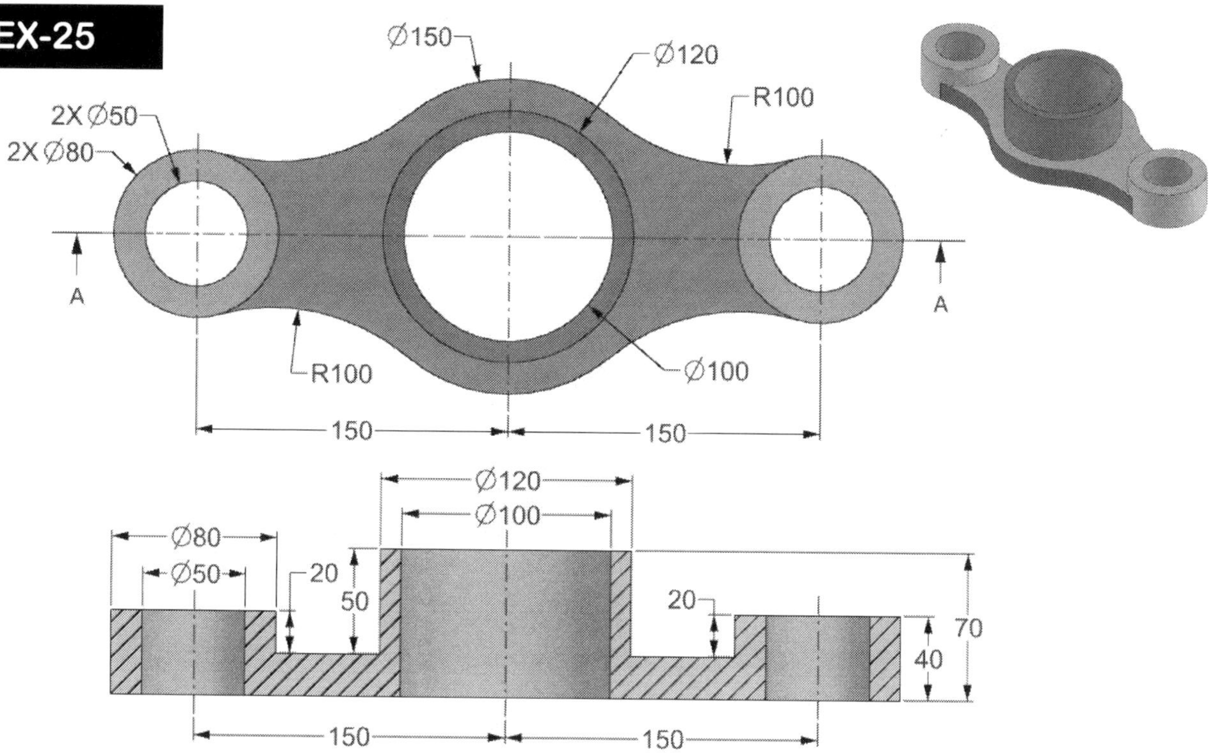

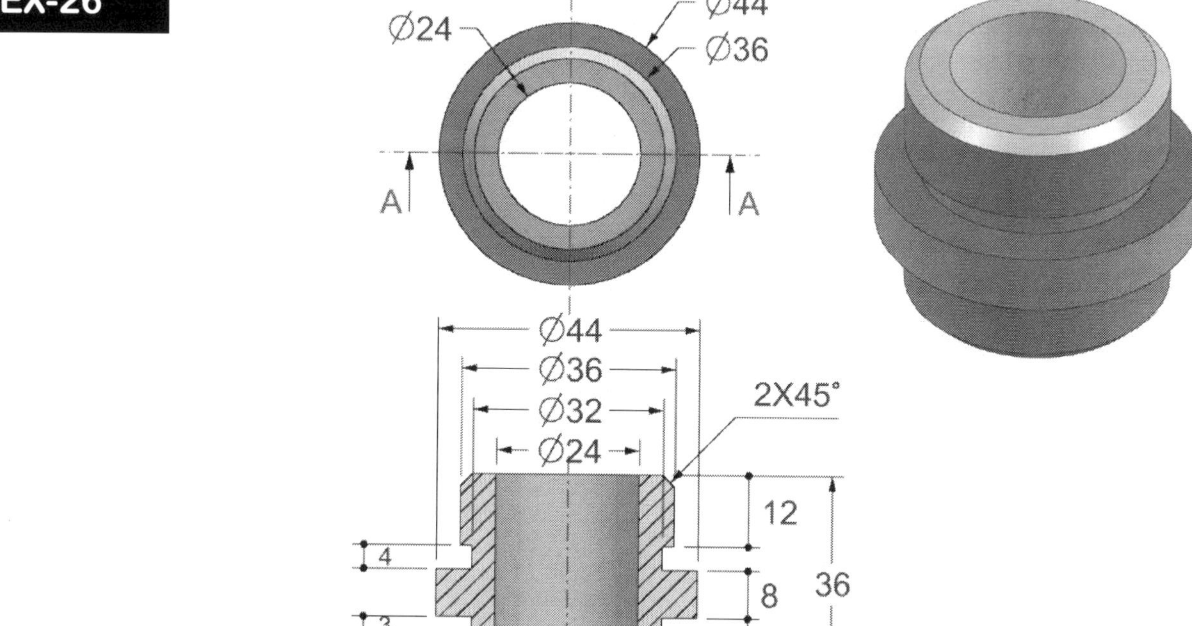
Ø150
Ø120
R100
2X Ø50
2X Ø80
Ø100
R100
A
A
150
150
Ø80
Ø50
Ø120
Ø100
20
50
20
70
40
150
150
SECTION A-A
(SCALE 1:1)
Ø24
Ø44
Ø36
A
A
Ø44
Ø36
Ø32
Ø24
2X45°
4
3
12
8
12
36
SECTION A-A
(SCALE 1:1)

EX-27 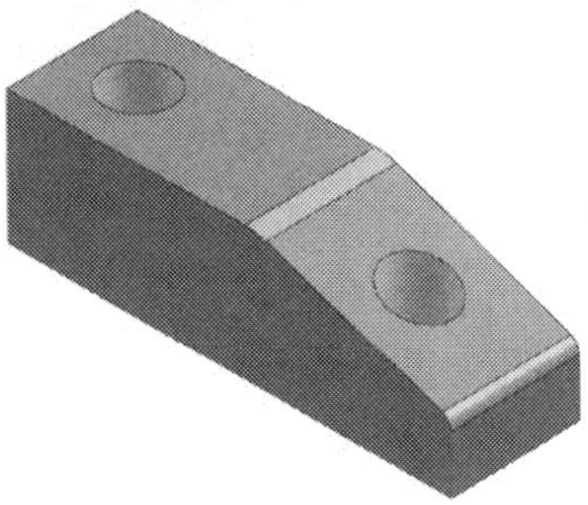 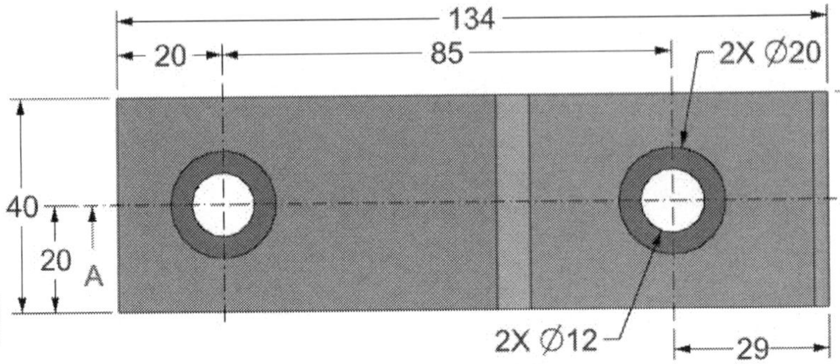 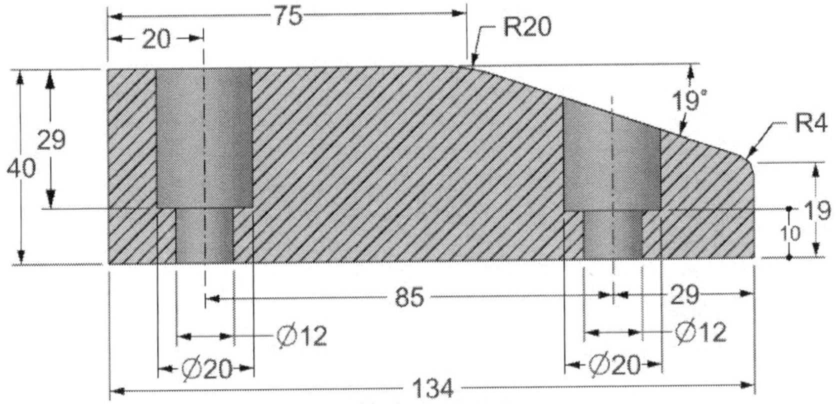
134
20
85
2X Ø20
40
20
20
20
A
A
2X Ø12
29
75
R20
20
R4
29
19°
40
19
10
85
29
Ø12
Ø12
Ø20
Ø20
134
SECTION A-A
(SCALE 1:1)
EX-28 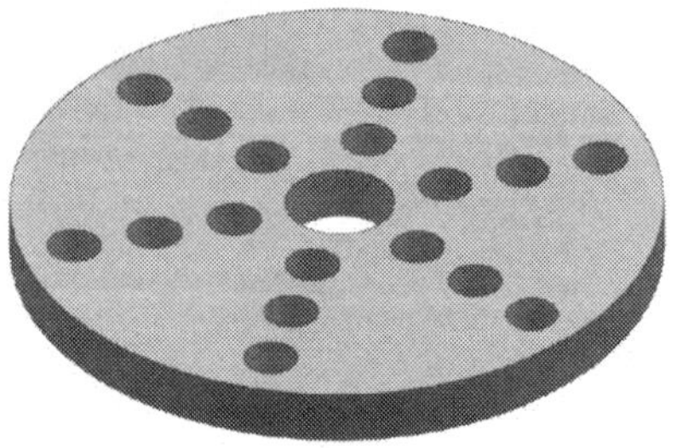 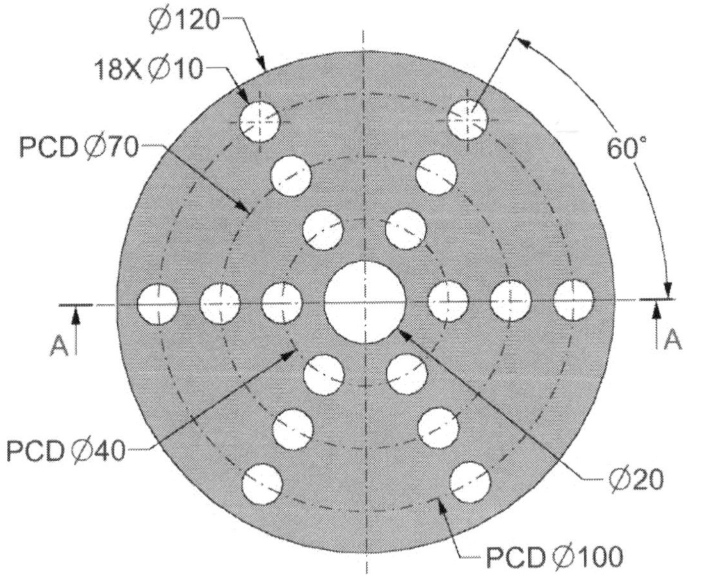 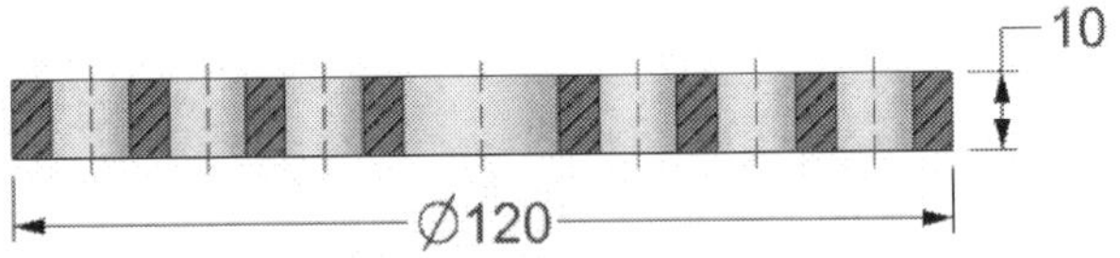
Ø120
18X Ø10
PCD Ø70
60°
PCD Ø40
Ø20
A
A
PCD Ø100
10
Ø120
SECTION A-A
P-14

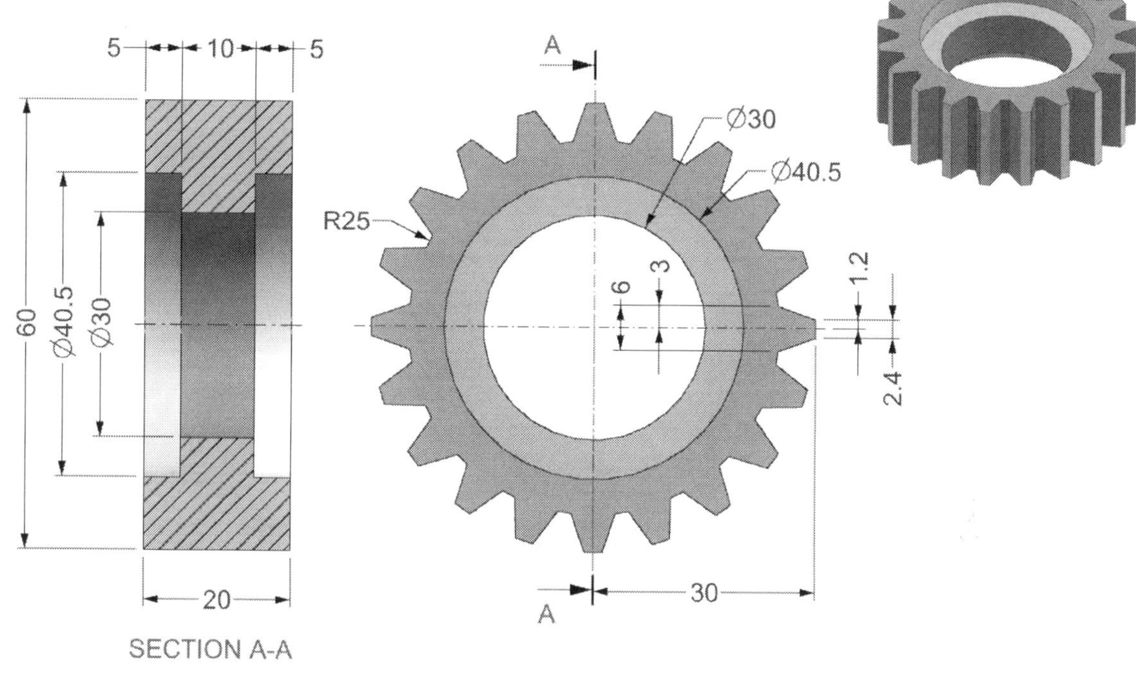

EX-29
5
10
5
60
Ø40.5
Ø30
20
SECTION A-A
A
Ø30
Ø40.5
R25
6
3
1.2
2.4
30
A
EX-30
90
70
10
10
50
10
10
30
10
10
5
Ø5
Ø10
10
20
40
5
10
20
10
Ø50
Ø10
Ø10
25
40
20
Ø5
10
10
10
10
20
10
10
10
10
10
10
30
10
10
70
10
10
90
25
40
10
10
5
10
10
20
10
40
10
10
5
5
P-15

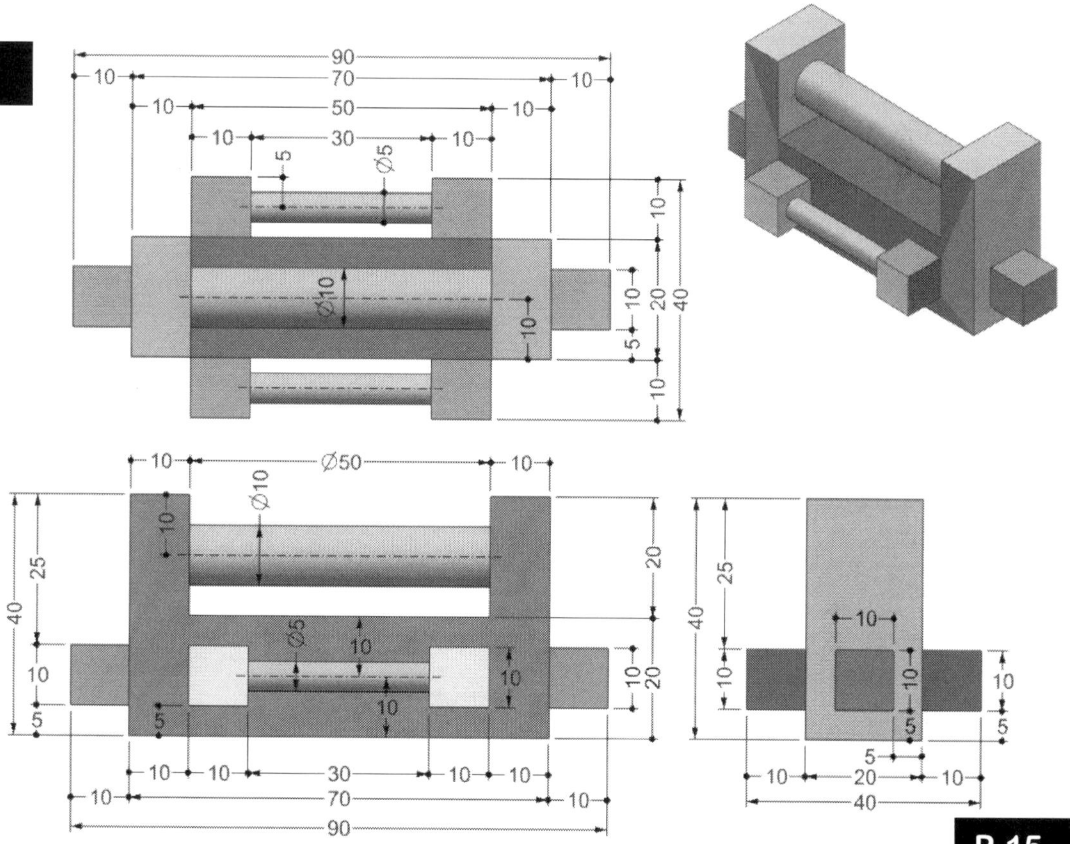

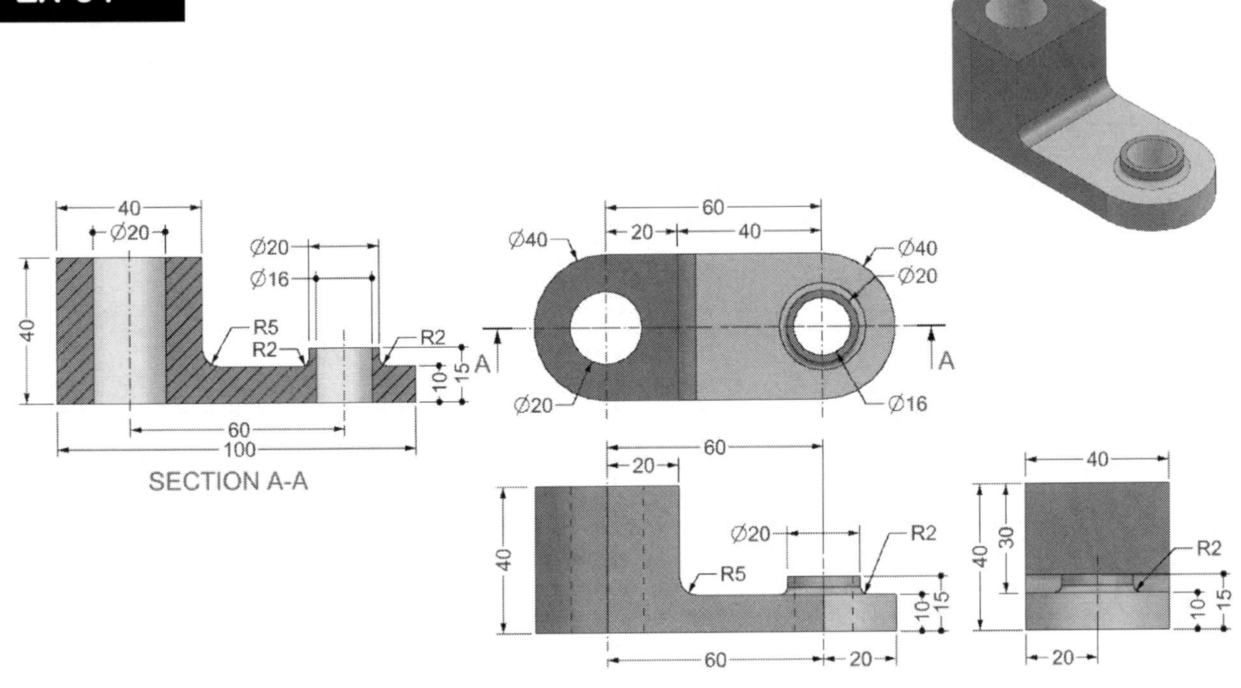
40
Ø20
Ø20
Ø16
R5
R2
R2
40
10
15
A
60
100
SECTION A-A
Ø40
20
40
60
Ø40
Ø20
Ø20
Ø16
A
60
20
Ø20
R2
R5
40
10
15
60
20
40
30
40
R2
10
15
20

3X Ø30
3X Ø60
40
100
70
100
100
Ø30 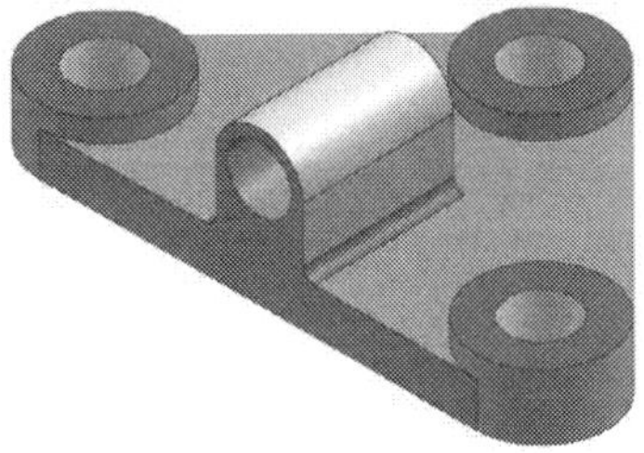 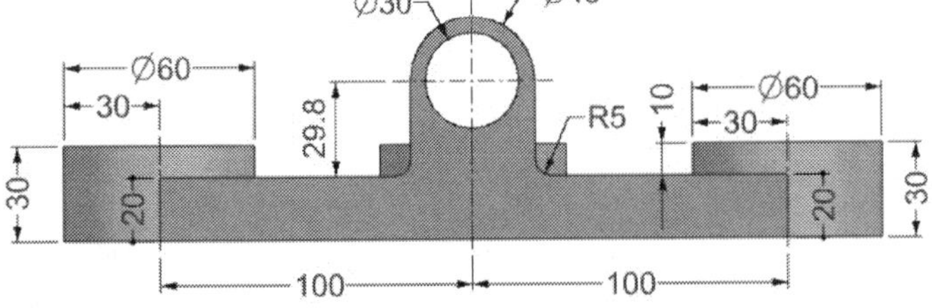
Ø40
Ø60
30
29.8
R5
10
Ø60
30
30
20
20
100
100
70
69.8
49.8
Ø60
30
Ø60
20
30
100

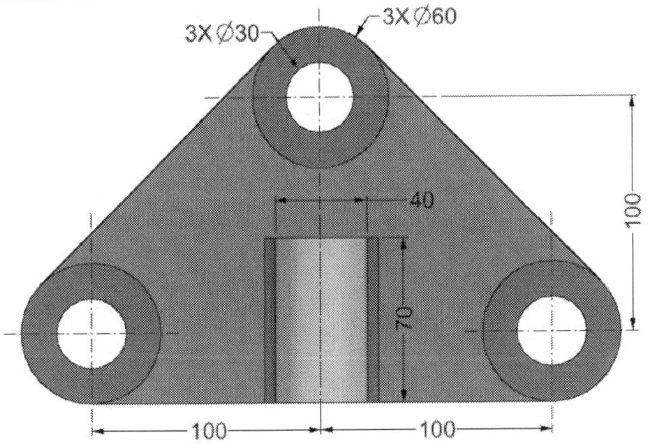

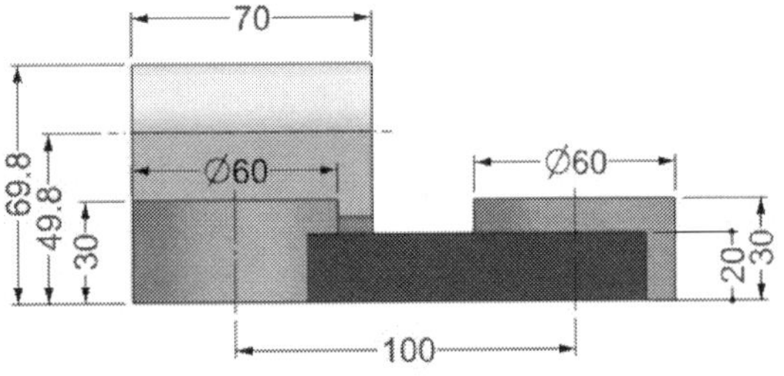

EX-33
3X Ø44
3X R35
112.5
30
15
65
65
130
Ø44
Ø44
20
130
SECTION A-A
A
A

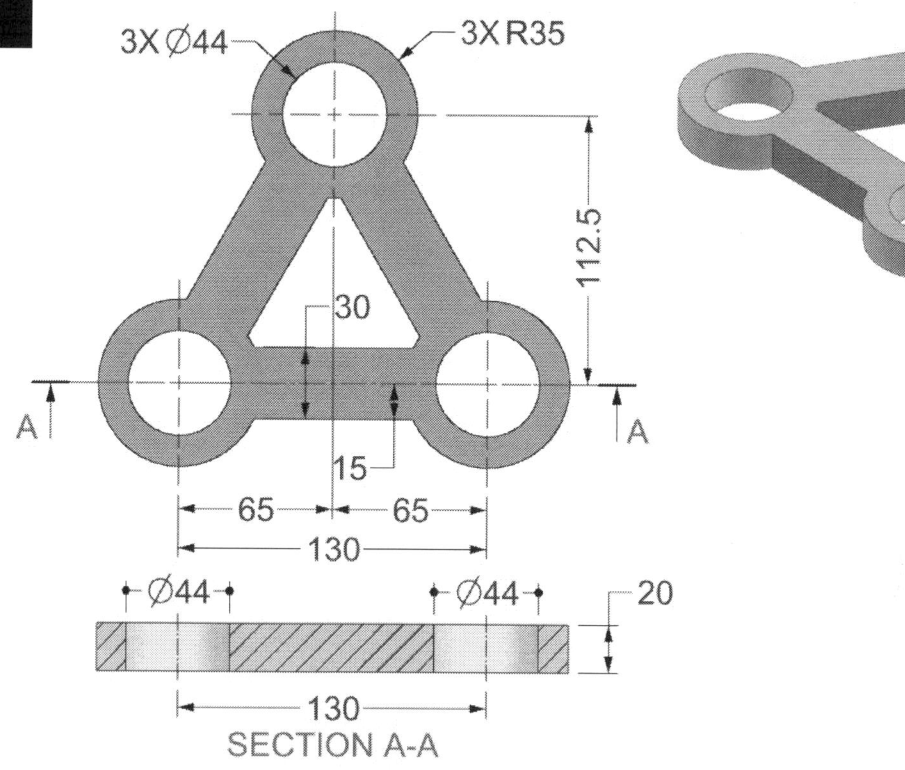

EX-34
20
80
15 x 45°
30
45
120
Ø60
15
15
R15
15
45
30
20
45
35
20
20
100
80
R10
15
100
60
20 x 45°
R30
23
20
2X ⌴ Ø16 ↧8
2X Ø12
THRU ALL
74
20
Ø30
33
100
53
15 30 15 15 30 15
120
20
20

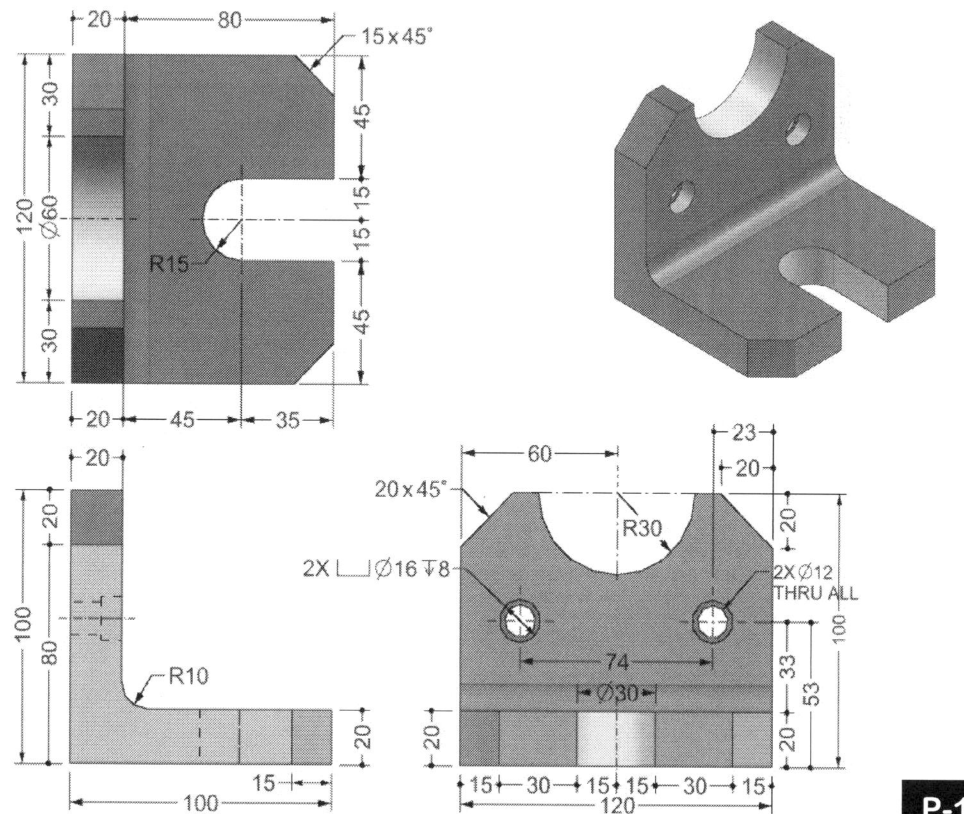

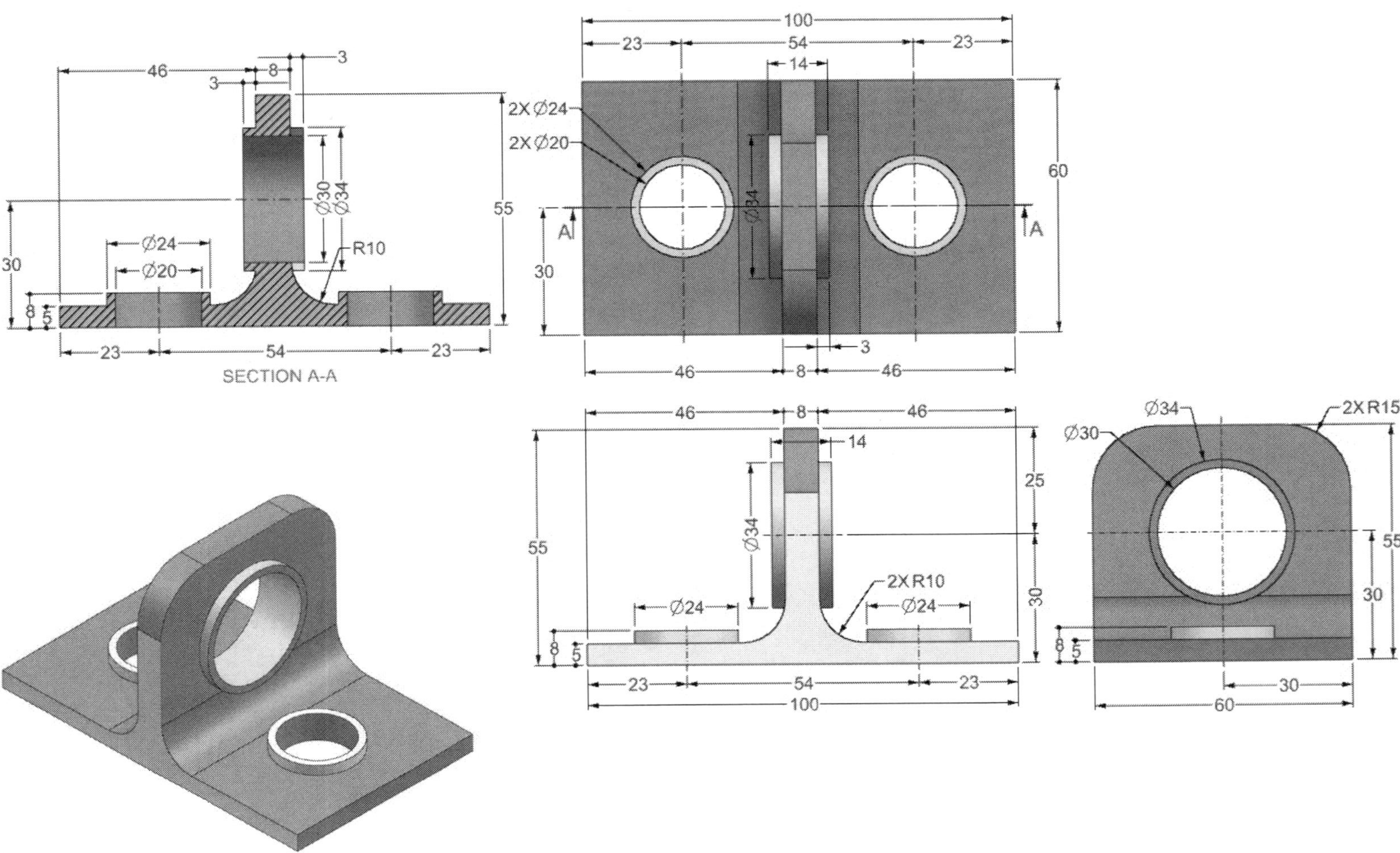

46
8
3
3
2X Ø24
2X Ø20
Ø30
Ø34
55
Ø24
Ø20
R10
30
8
5
23
54
23
SECTION A-A
100
23
54
23
14
60
Ø34
30
A
A
46
8
3
46
46
8
46
14
25
Ø34
55
2X R10
Ø24
Ø24
30
8
5
23
54
23
100
Ø34
Ø30
2X R15
55
30
8
5
30
60

EX-36
6X Ø10
60°
60°
PCD Ø90
Ø45
Ø120
Ø120
PCD 90
Ø45
2 X 45°
2 X 45°
20
Ø10
R3
60
2 X 45°
2.5
Ø45
Ø50
(SCALE 1:1) SECTION A-A
A
A

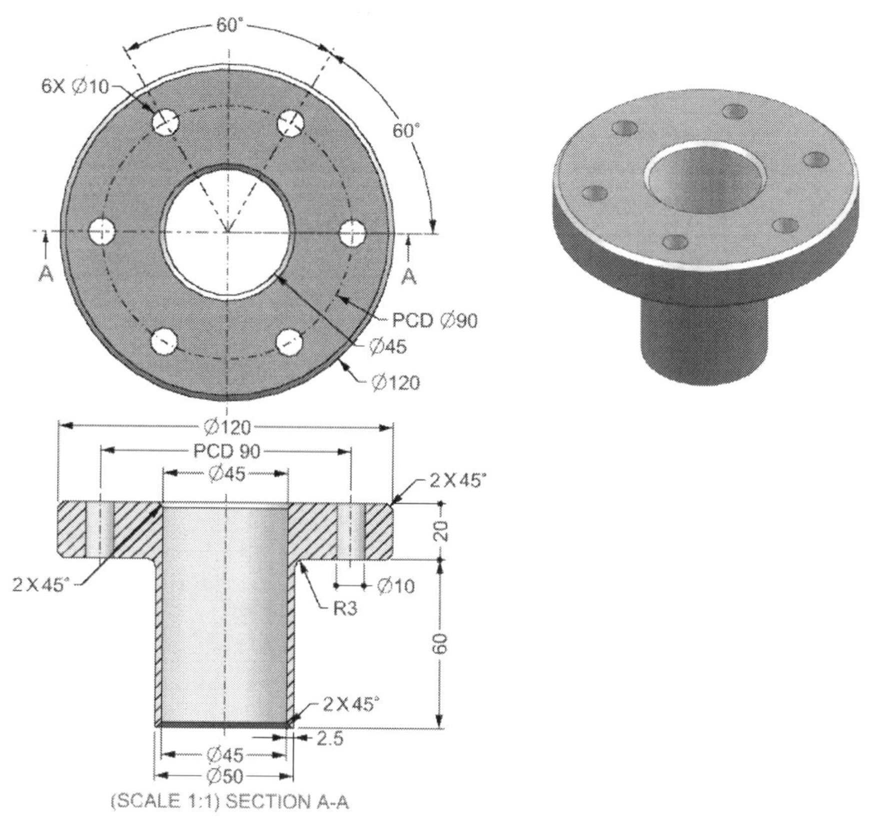

EX-37 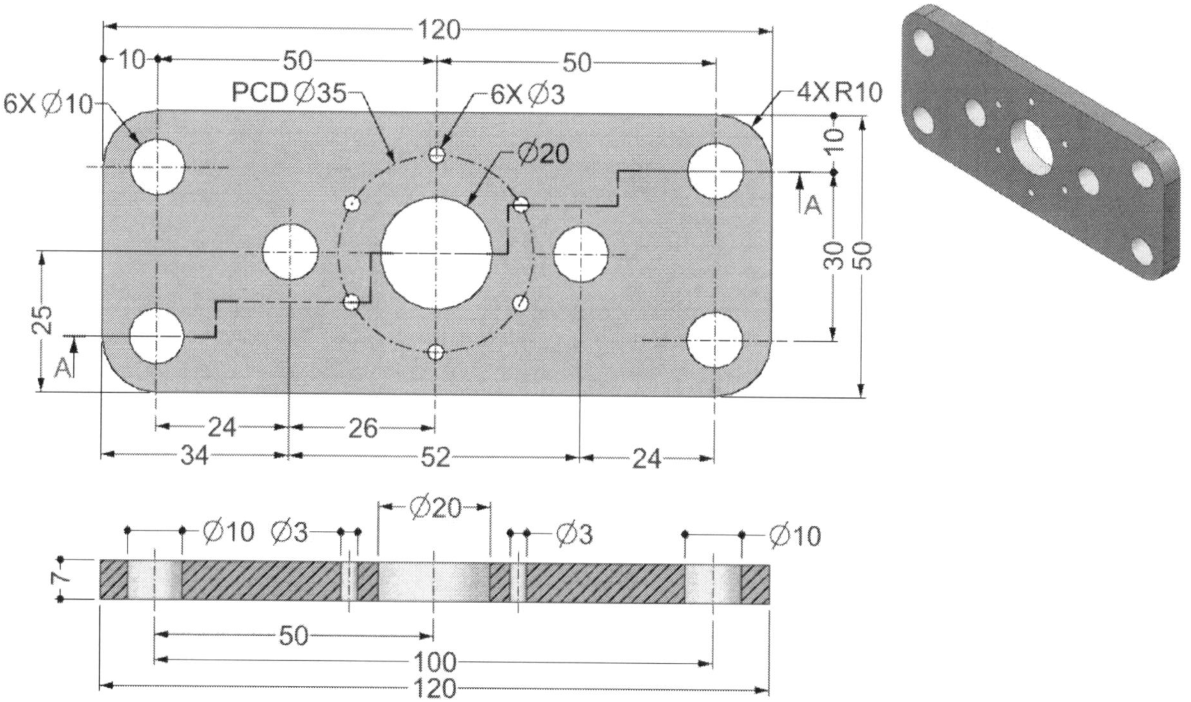
120
10
50
50
6X Ø10
PCD Ø35
6X Ø3
4X R10
Ø20
10
A
30
50
25
A
24
26
34
52
24
Ø10 Ø3
Ø20
Ø3
Ø10
7
50
100
120
SECTION A-A
A
A

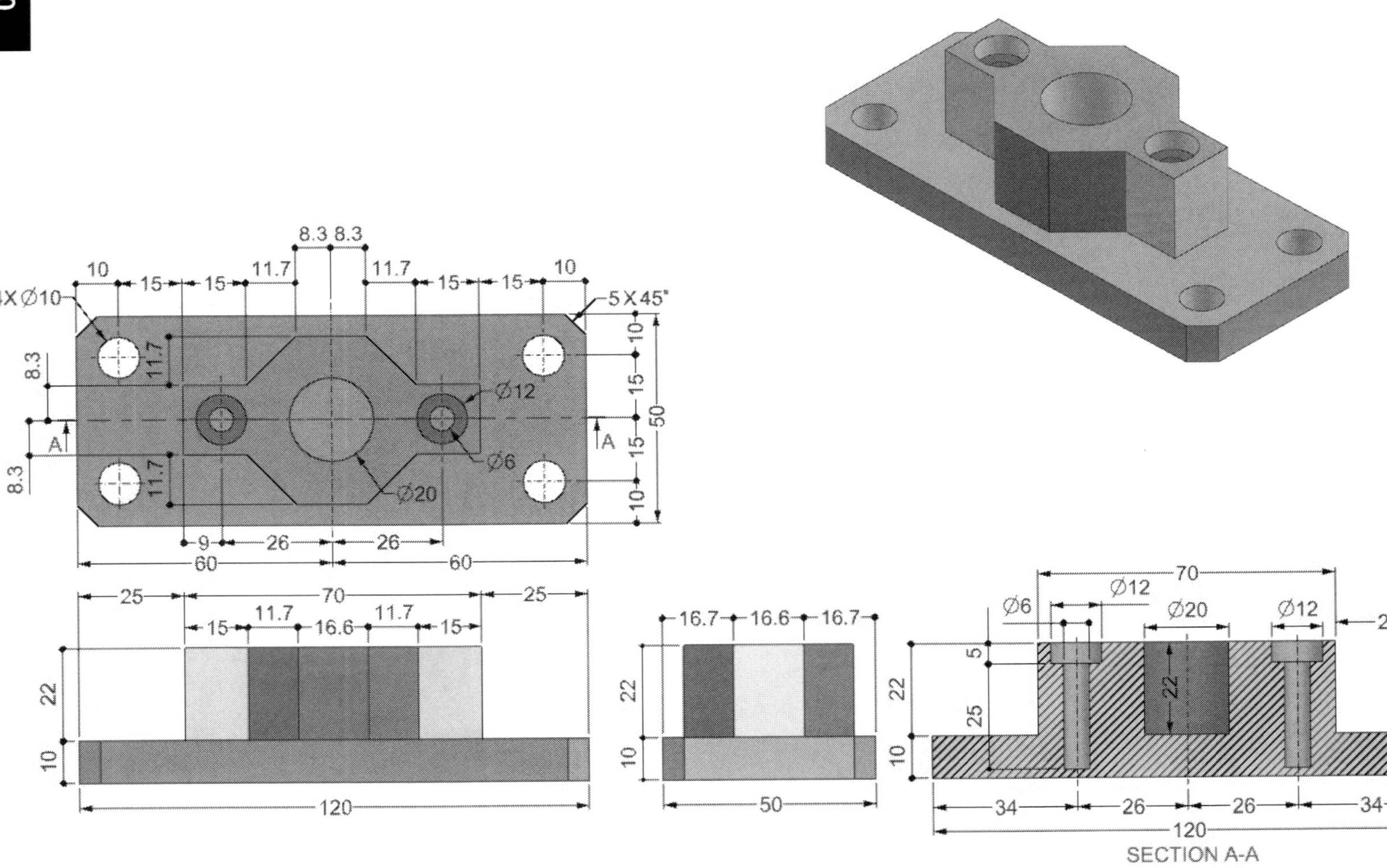

4X Ø10
8.3 8.3
5 X 45°
Ø12
Ø6
Ø20
10
15
15
11.7
11.7
15
15
10
8.3
8.3
11.7
9
26
26
60
60
50
25
11.7
70
11.7
25
16.7
16.6
16.7
Ø6
Ø12
Ø20
Ø12
70
22
25
22
22
10
10
10
120
50
34
26
26
34
120
SECTION A-A

70
R20
Ø20
40
45

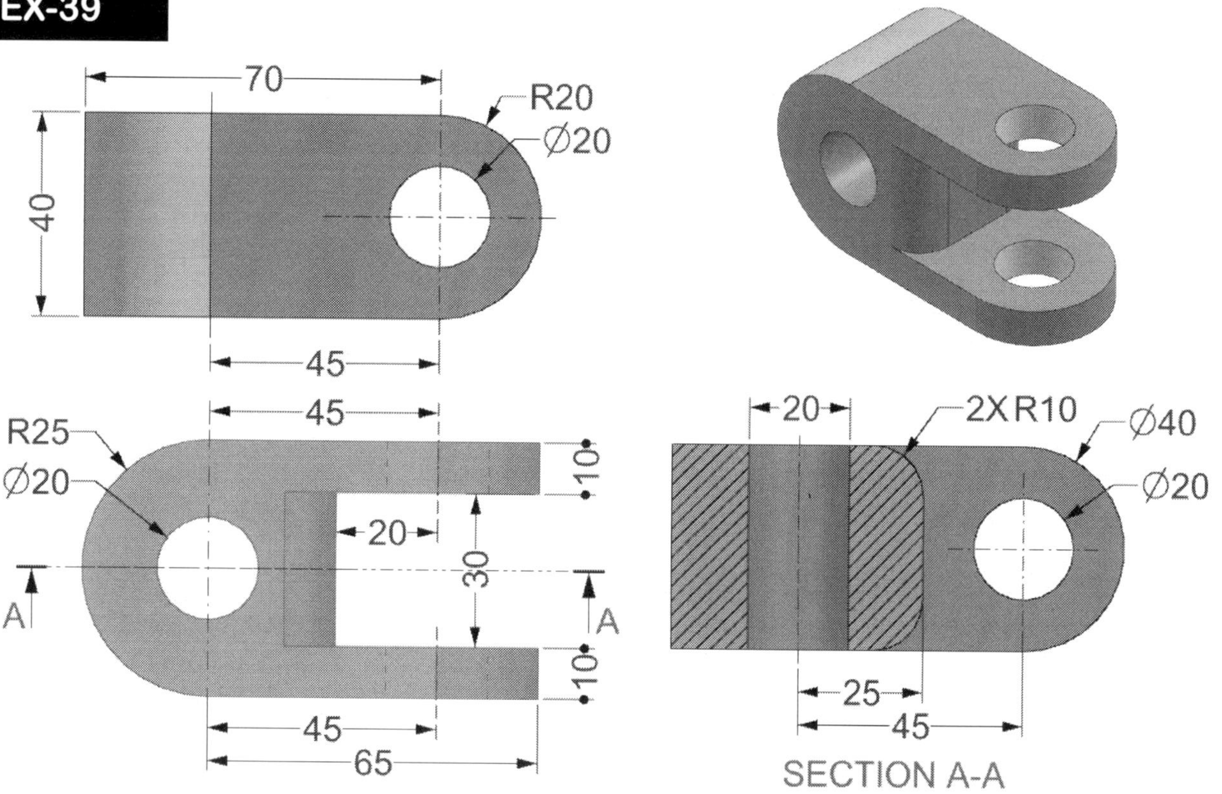

R25
Ø20
45
20
30
10
10
45
65
A
A

20
2X R10
Ø40
Ø20
25
45

SECTION A-A

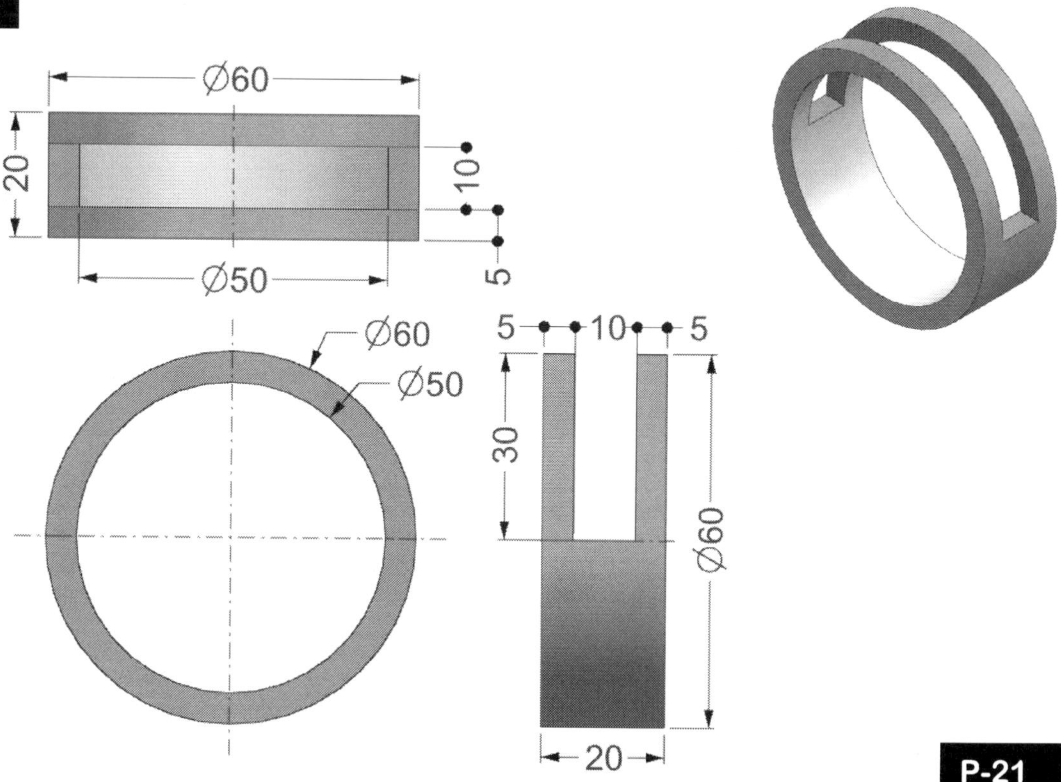

Ø60
20
10
5
Ø50

Ø60
Ø50

5
10
5
30
Ø60
20

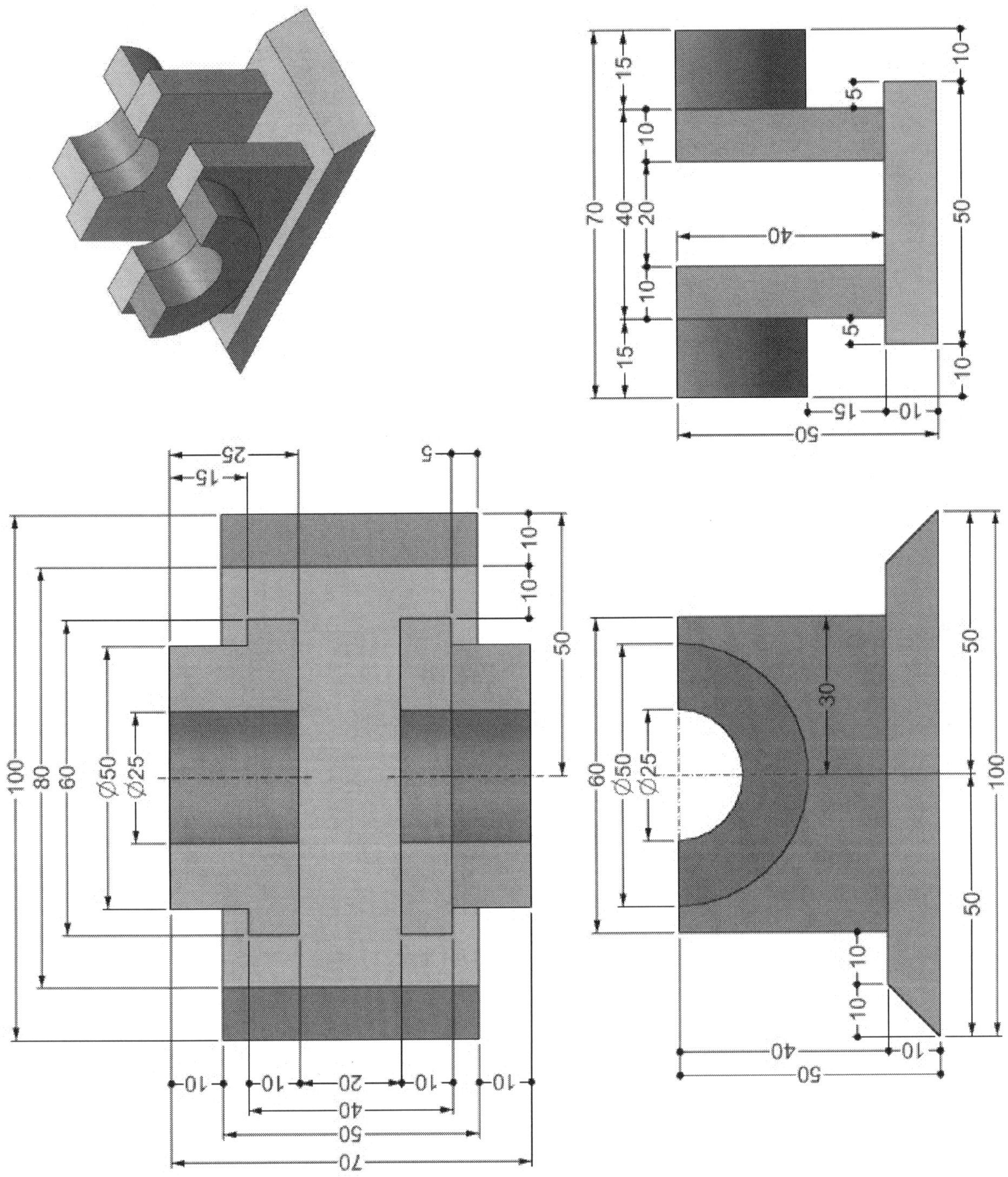

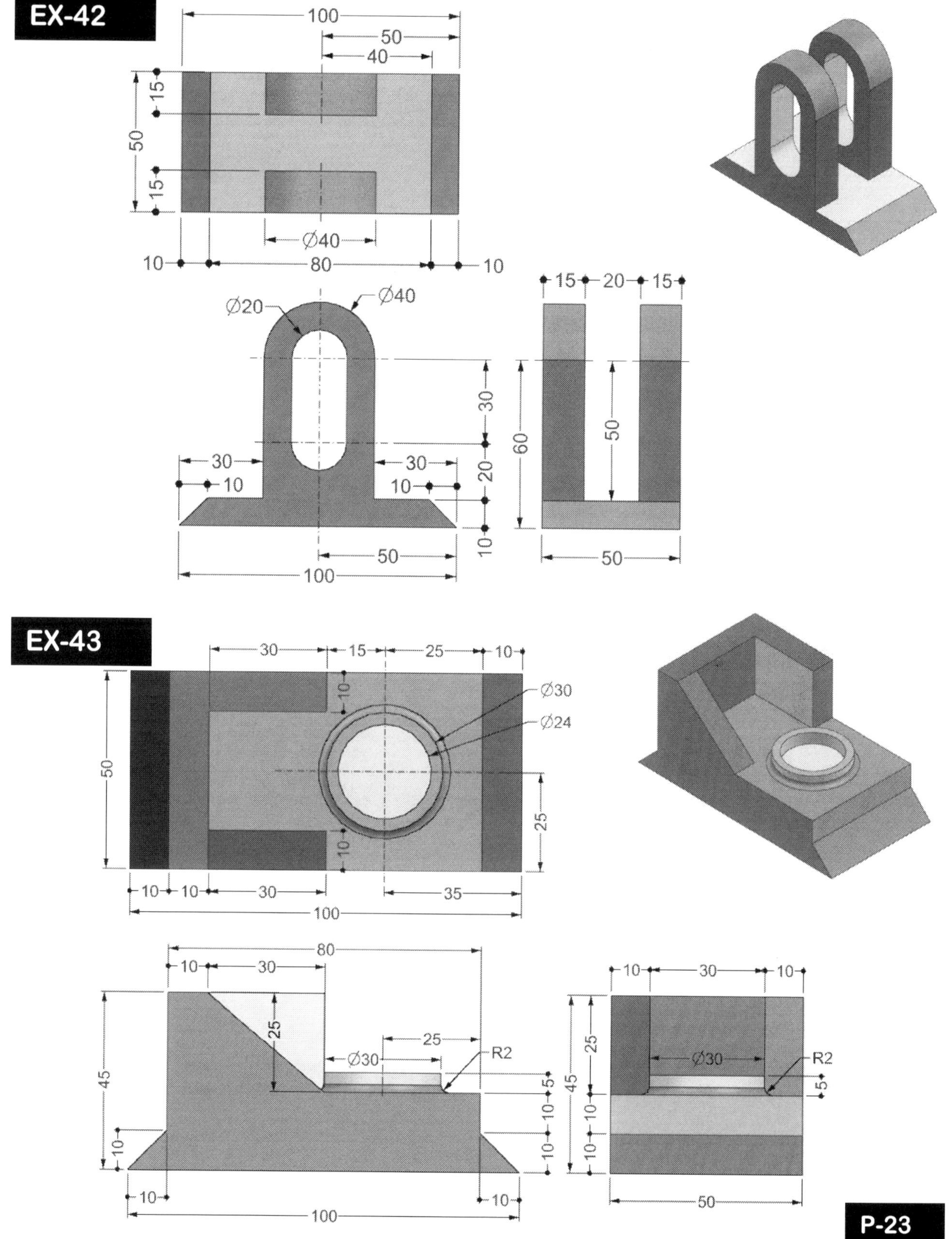

EX-42
100
50
40
15
50
15
Ø40
80
10
10
Ø20
Ø40
30
30
10
10
60
30
20
10
50
100
15
20
15
50
50

EX-43
30
15
25
10
10
Ø30
Ø24
50
25
10
10
10
30
35
100
80
10
30
25
25
Ø30
R2
45
10
5
10
10
10
100
10
30
10
25
45
Ø30
R2
5
10
10
50

P-23

EX-44

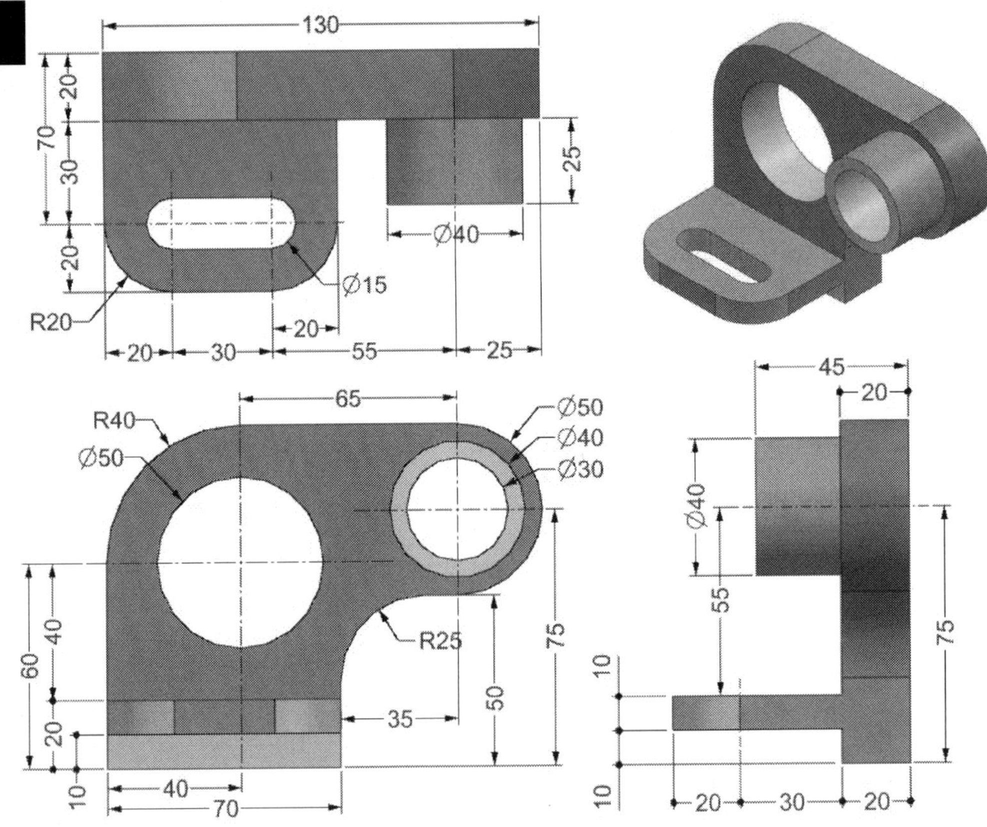

130
20
70
30
25
20
Ø40
Ø15
R20
20
30
20
55
25
R40
65
Ø50
Ø40
Ø30
Ø50
60
40
R25
75
50
20
35
10
40
70
45
20
Ø40
55
75
10
10
20
30
20

EX-45

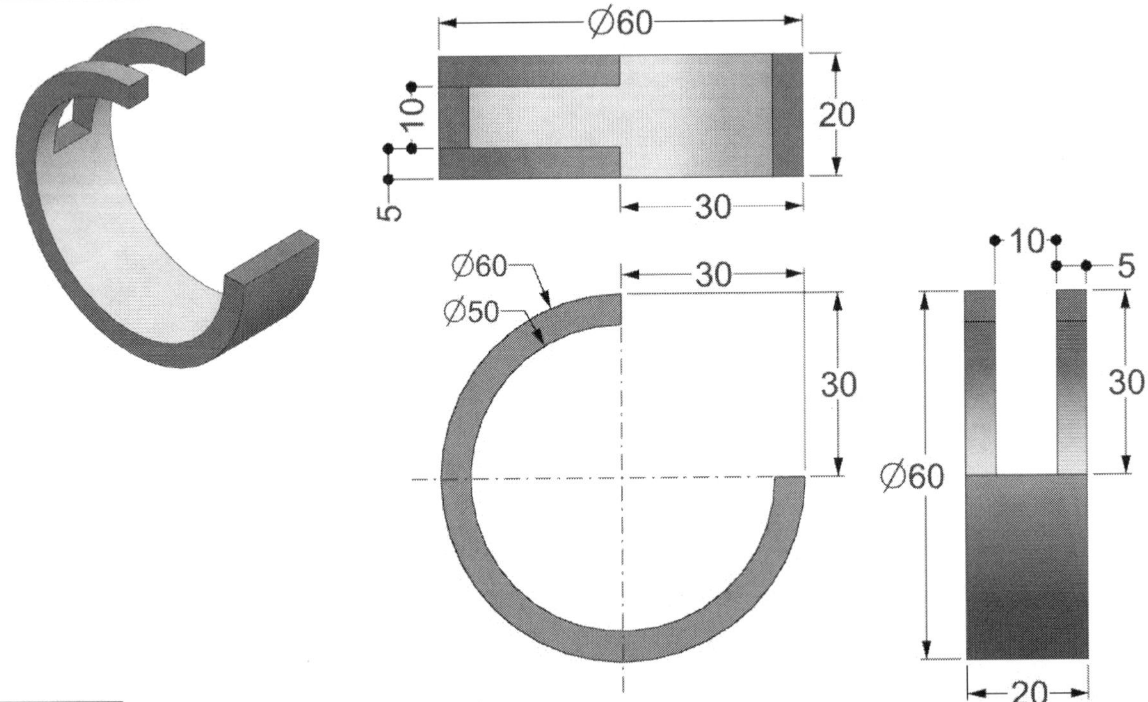

Ø60
10
20
5
30
Ø60
Ø50
30
30
30
Ø60
10
5
30
20

EX-46

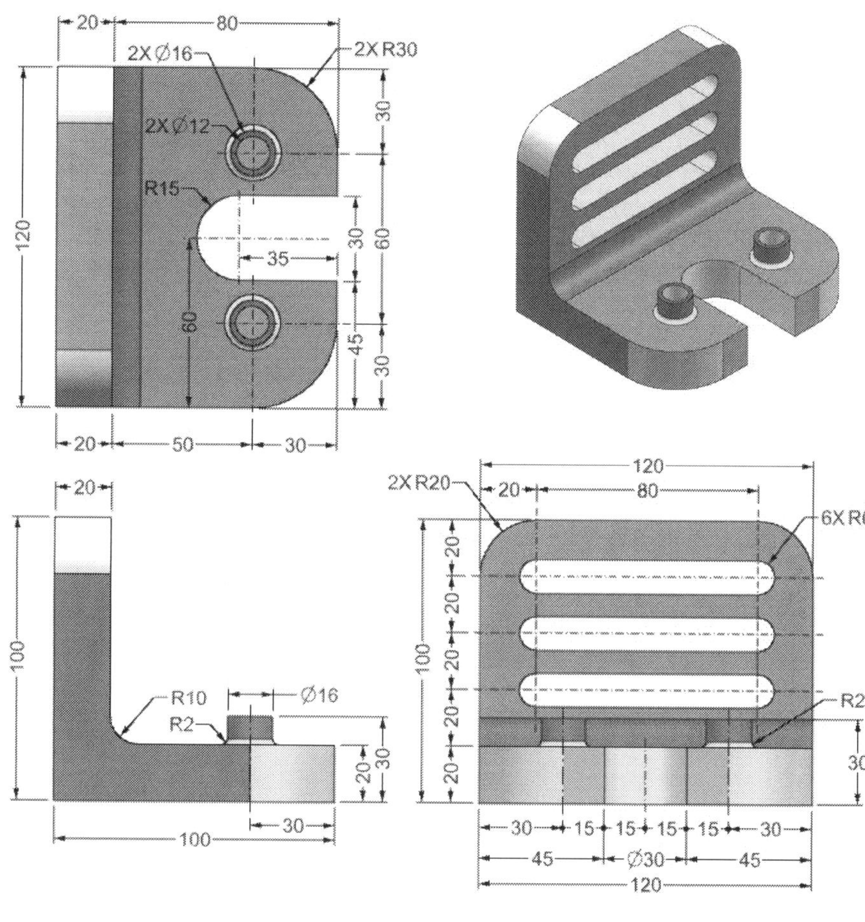
20
80
2X Ø16
2X R30
2X Ø12
30
R15
120
30
60
35
60
45
20
50
30
2X R20
20
120
80
6X R6
20
20
20
100
20
R2
20
30
30 15 15 15 15 30
45 Ø30 45
120
20
100
R10
Ø16
R2
20
30
100
30

EX-47

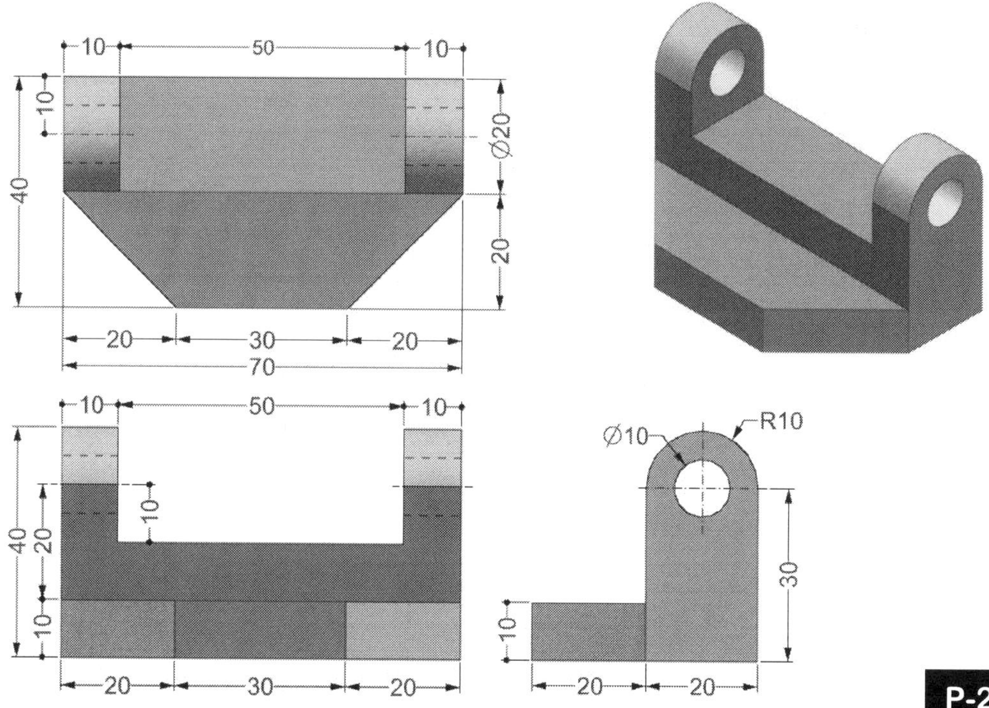
10
50
10
10
Ø20
40
20
20
30
20
70
10
50
10
40
10
20
10
20
30
20
Ø10
R10
30
10
20
20

P-25

EX-48

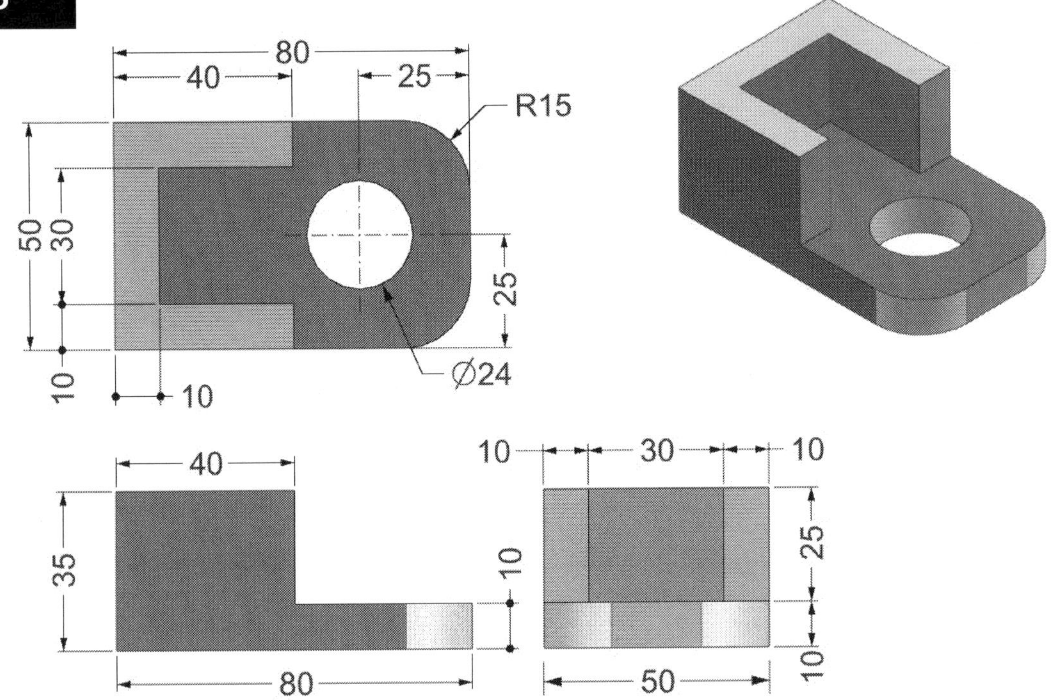
80
40
25
R15
50
30
25
10
10
Ø24
40
35
80
10
30
10
25
10
50

EX-49

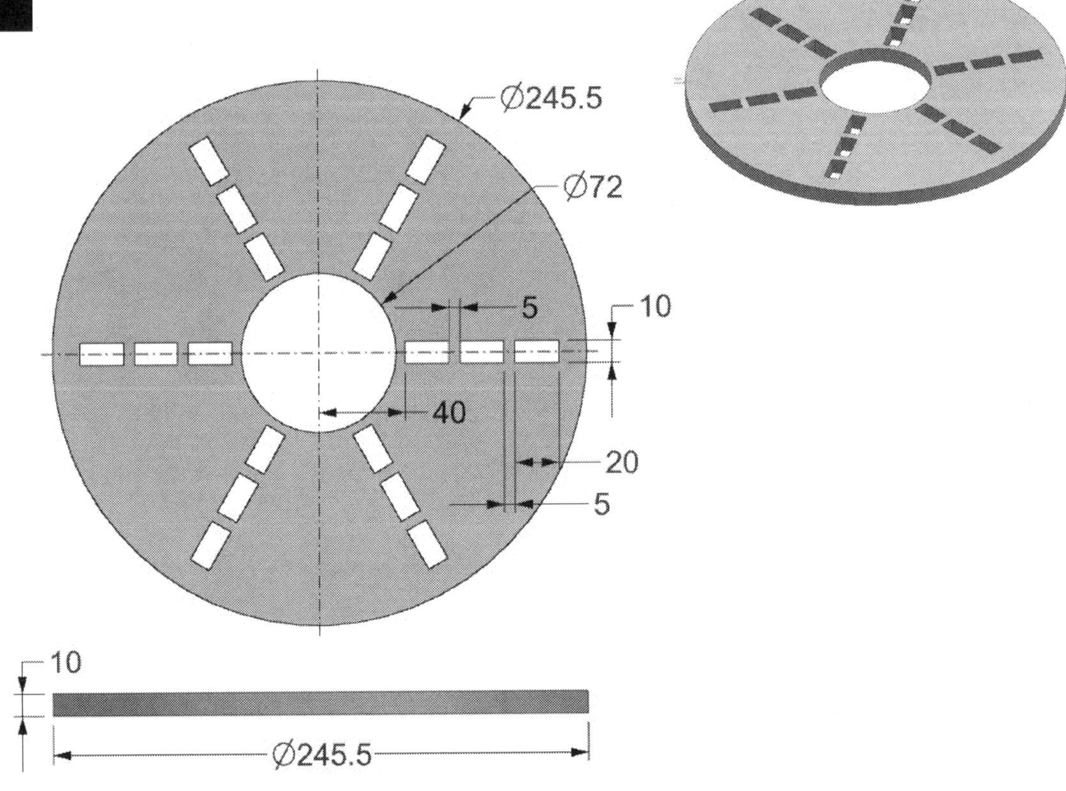
Ø245.5
Ø72
5
10
40
20
5
10
Ø245.5

P-26

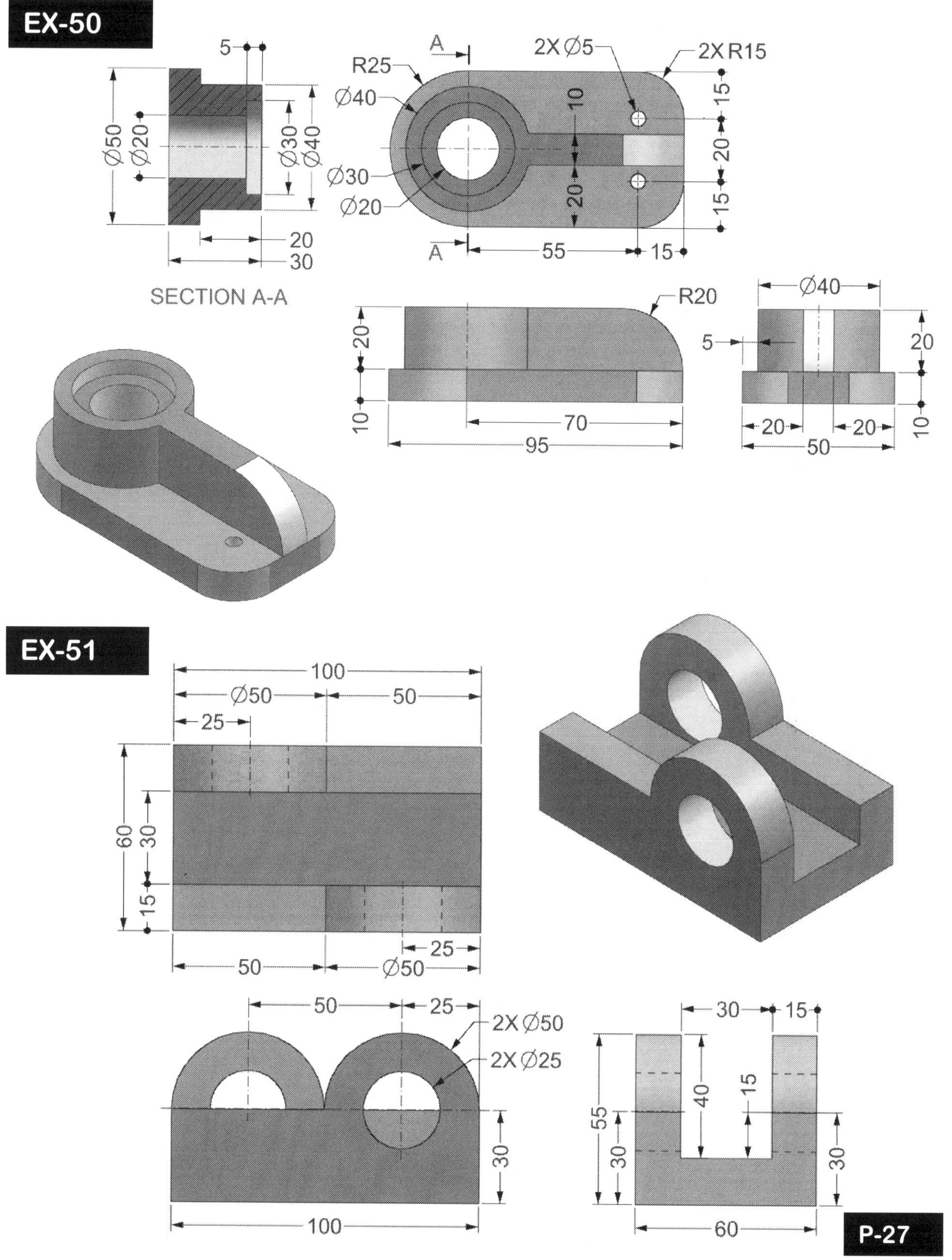

EX-52

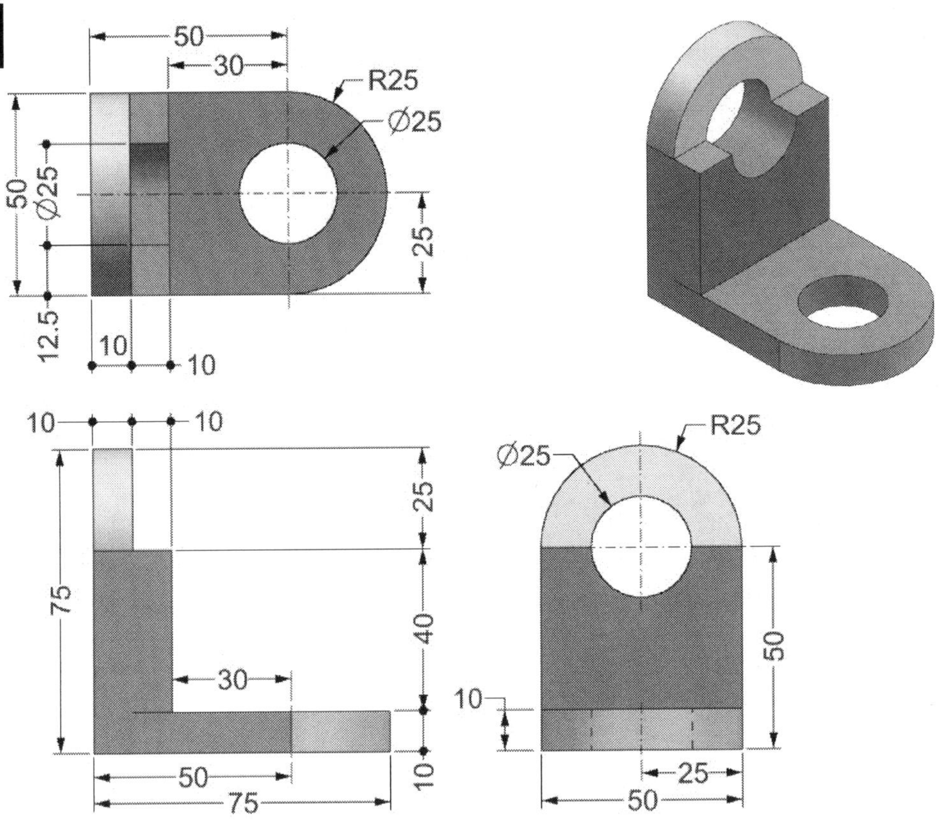

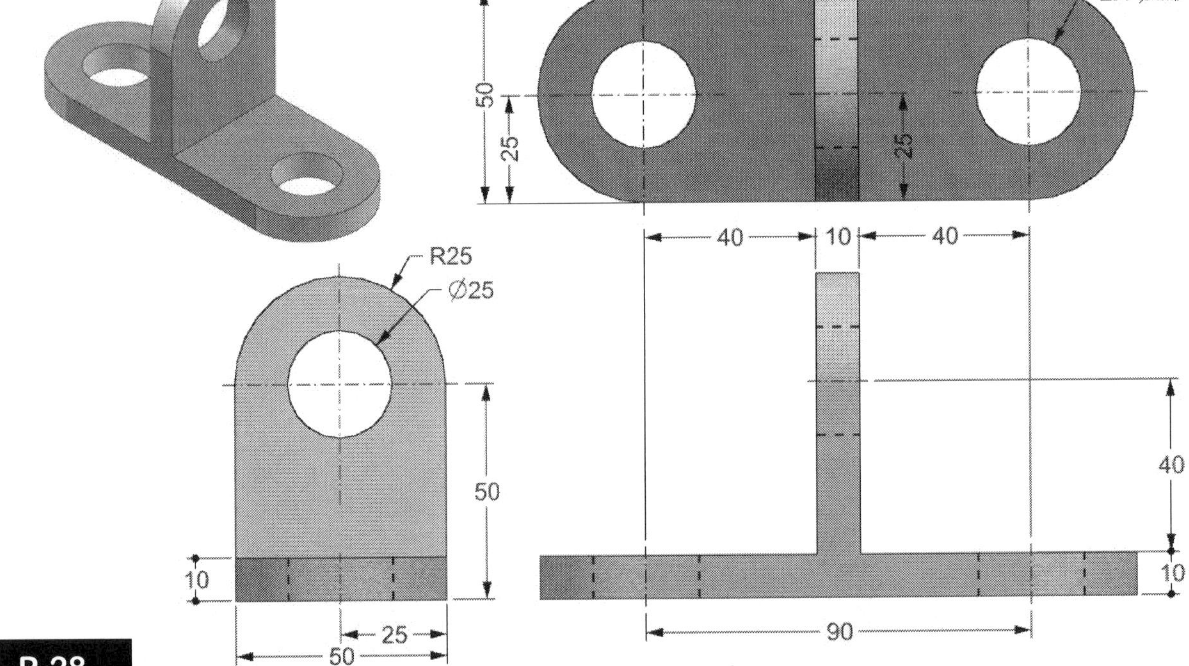

EX-52
50
30
R25
Ø25
50
Ø25
25
12.5
10
10
10
10
25
75
40
30
50
10
75
Ø25
R25
50
10
25
50
EX-53
90
40
40
2X R25
2X Ø25
50
25
25
40
10
40
R25
Ø25
50
10
25
50
40
10
90
P-28

EX-54
100 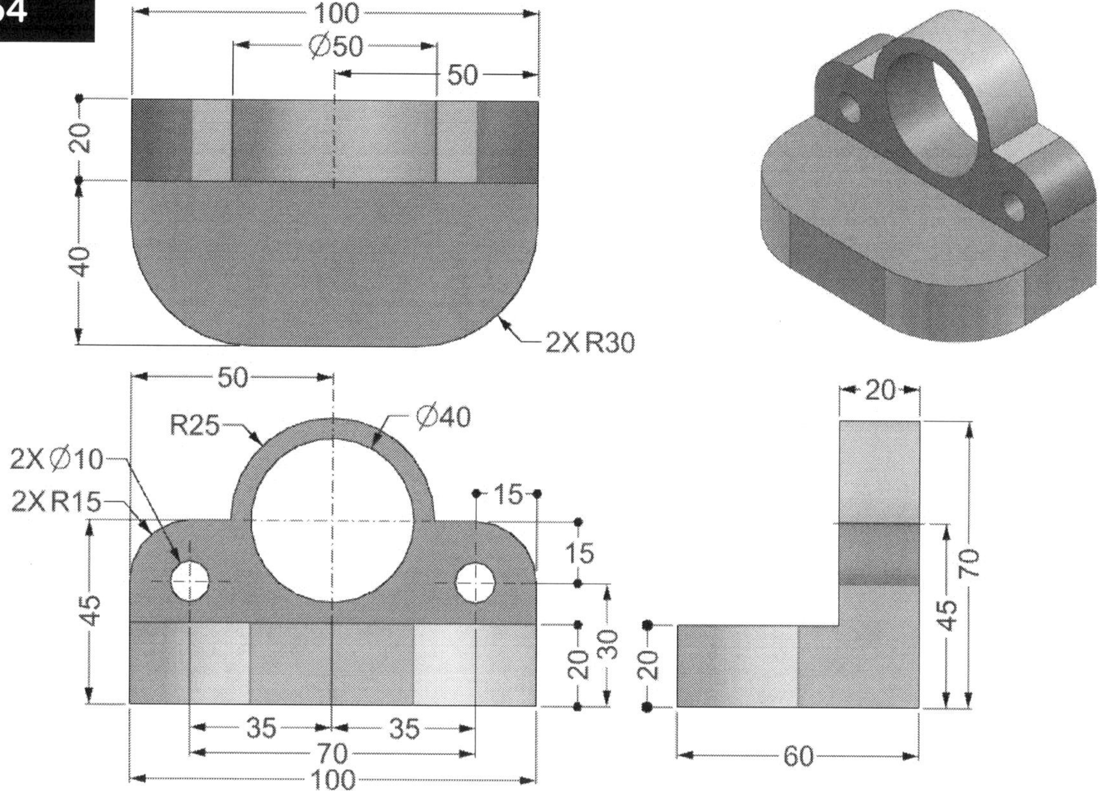
Ø50
50
20
40
2X R30
50
R25
Ø40
2X Ø10
2X R15
15
15
45
20
30
35
35
70
100
20
70
45
20
60
EX-55
100
20
60
50
35
35
20
60
20
50
15
35
35
15
50
50
P-29

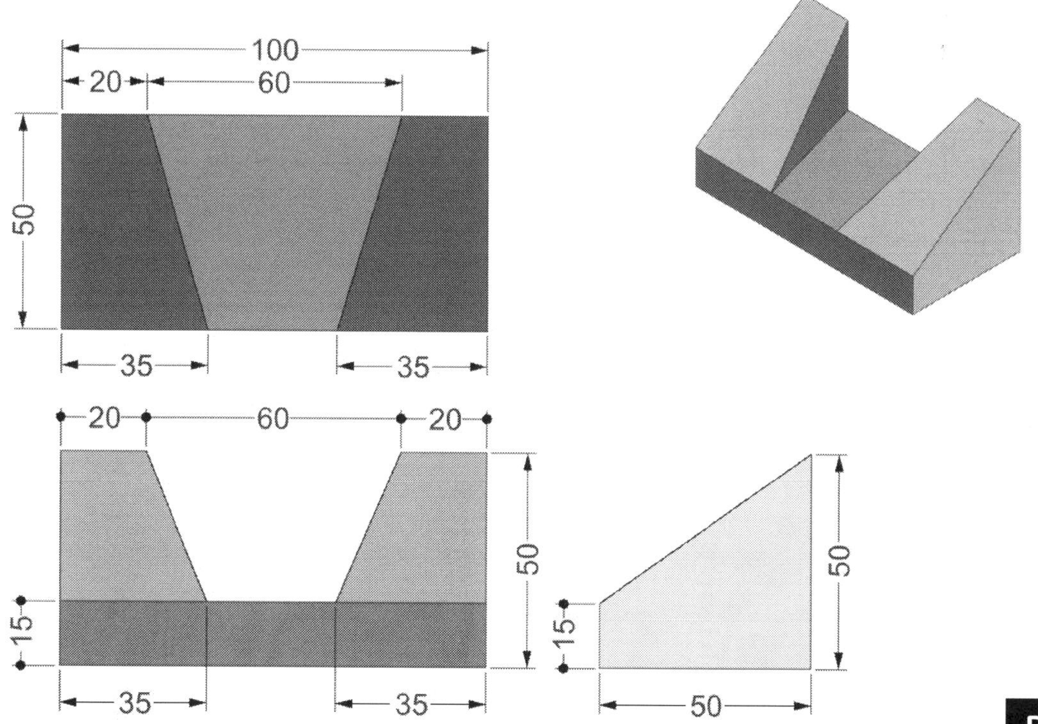

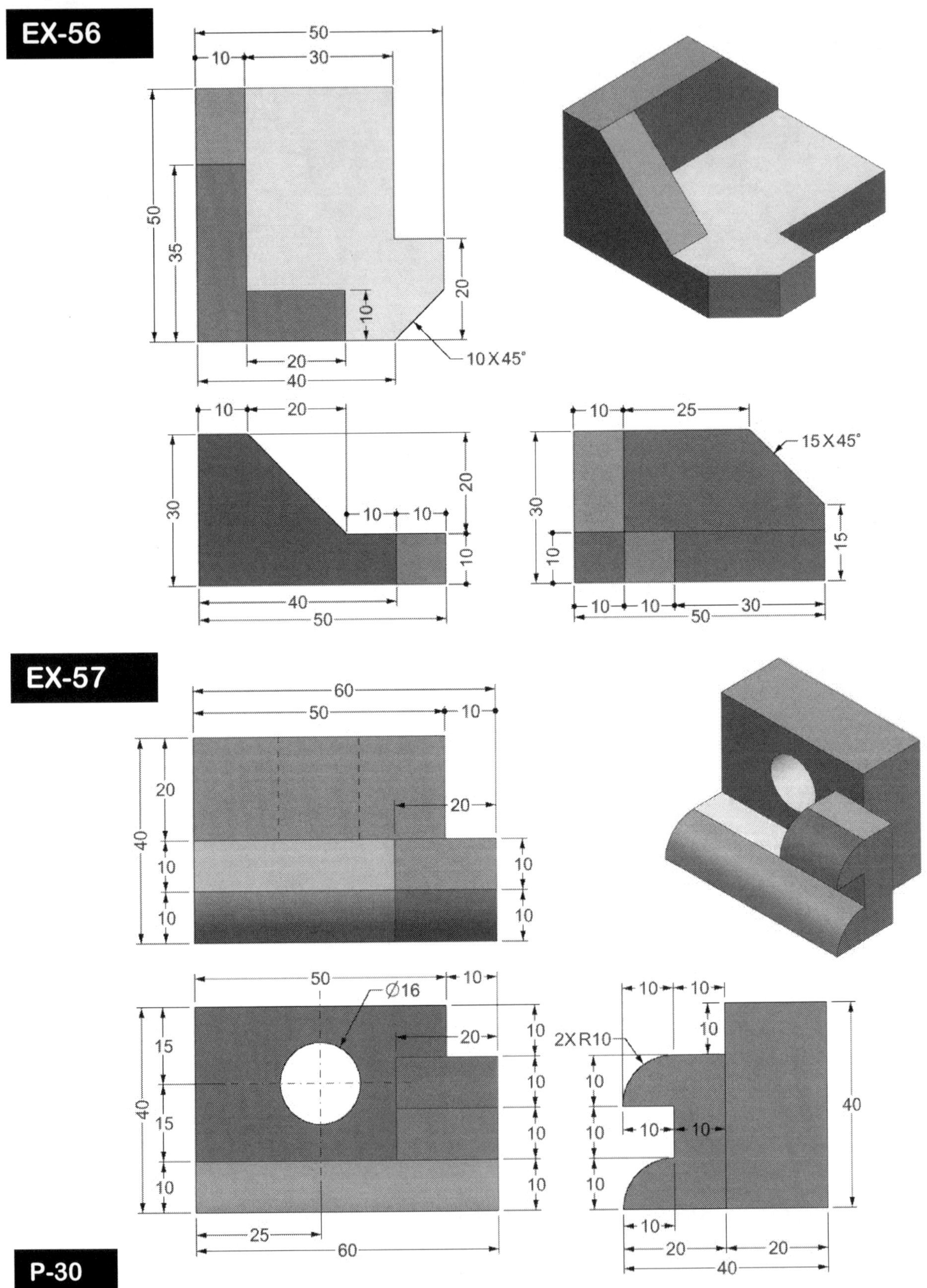

EX-56
50
10
30
50
35
20
10
20
40
10 X 45°
10
20
30
10
10
20
40
50
10
25
15 X 45°
30
10
15
10
10
50
30
EX-57
60
50
10
20
20
40
10
10
10
10
50
Ø16
10
20
10
15
2XR10
10
10
40
10
40
10
10
15
10
10
10
25
60
10
10
20
20
40
P-30

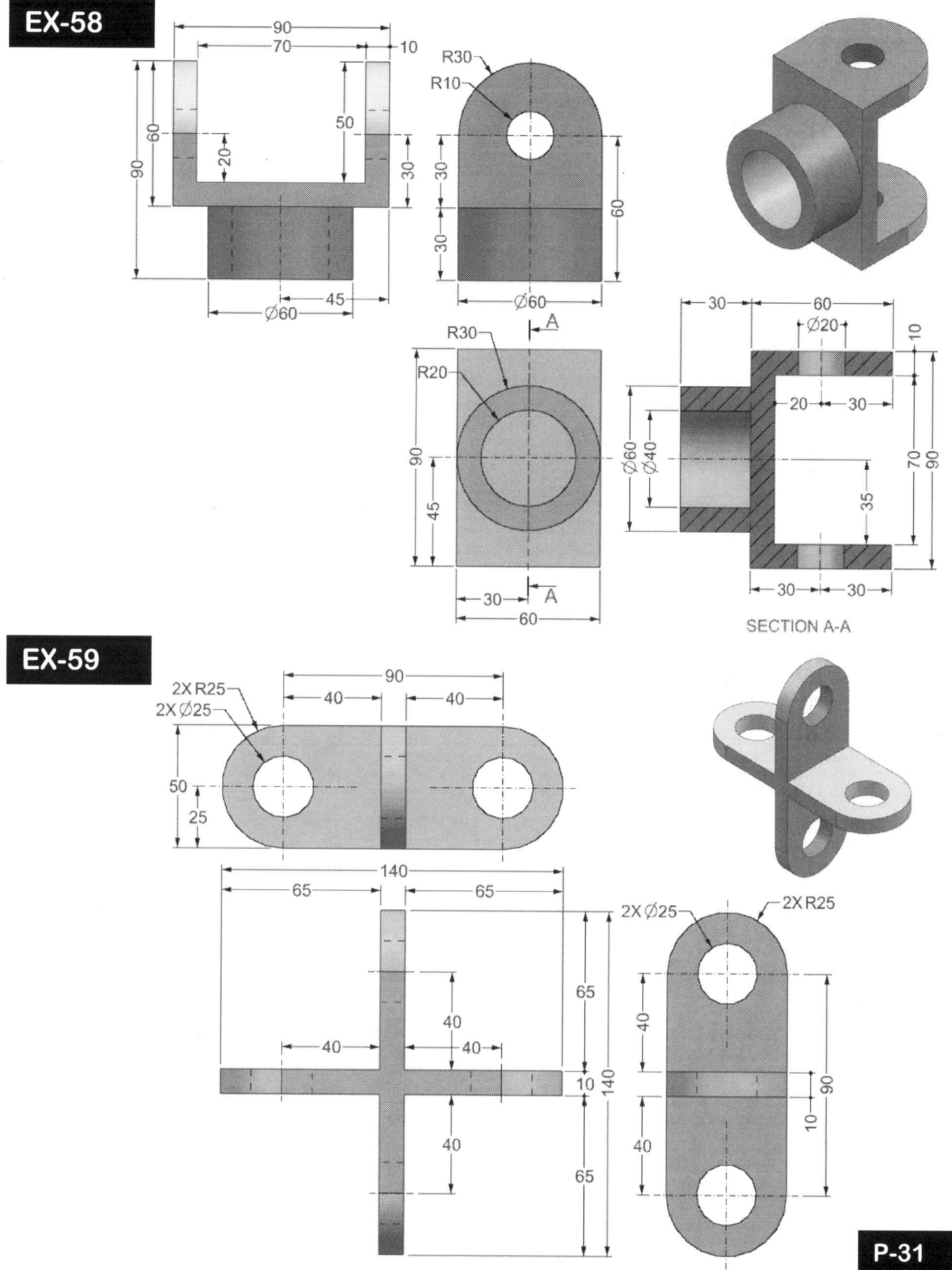

EX-58
90
70
10
R30
R10
60
50
60
90
20
30
30
30
Ø60
45
Ø60
Ø60
R30
R20
90
45
A
A
30
60
30
60
Ø20
10
20
30
Ø60
Ø40
70
90
35
30
30
SECTION A-A
EX-59
2X R25
2X Ø25
90
40
40
50
25
140
65
65
65
40
40
40
40
65
10
140
2X Ø25
2X R25
40
90
10
40
P-31

EX-60
Ø50
22.5
2X Ø10
15
25
43.9
10
60
15
25
50
45
95
Ø50
Ø40
R4
R10
R10
10
40
R10
R10
10
45
45
155
130
80
140
R10
R10
55
40
30
10
10
22.5
22.5
100
60
85.4
100
155
34.6
10
25
40
60 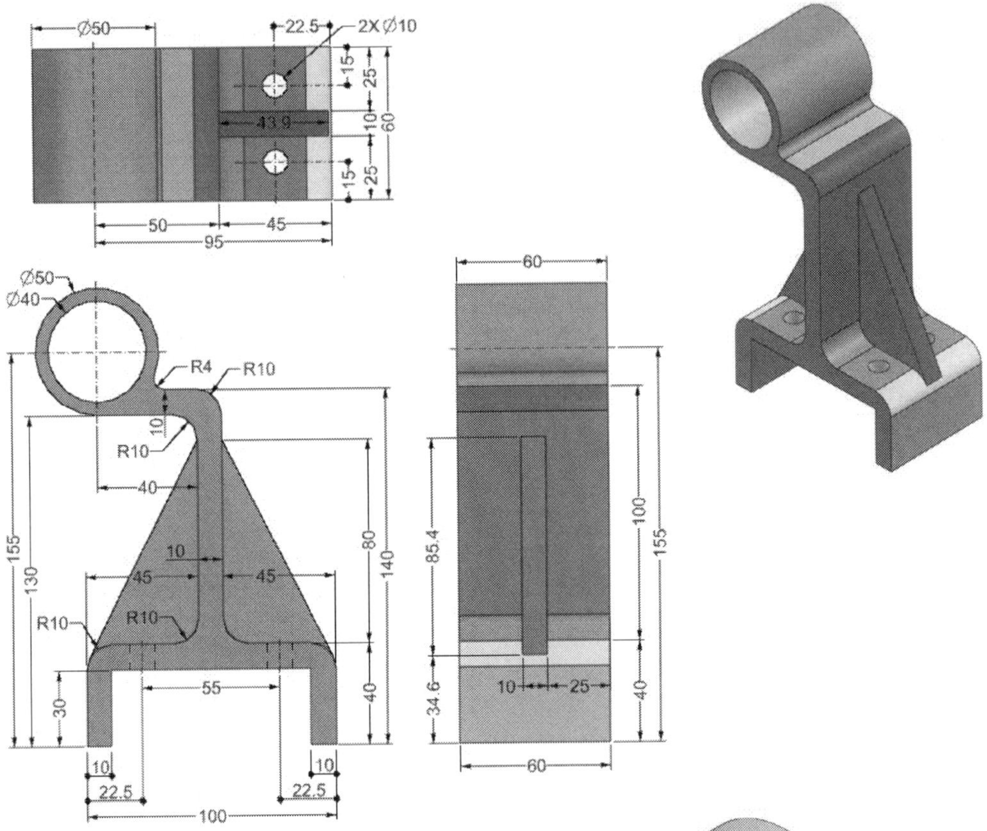
EX-61 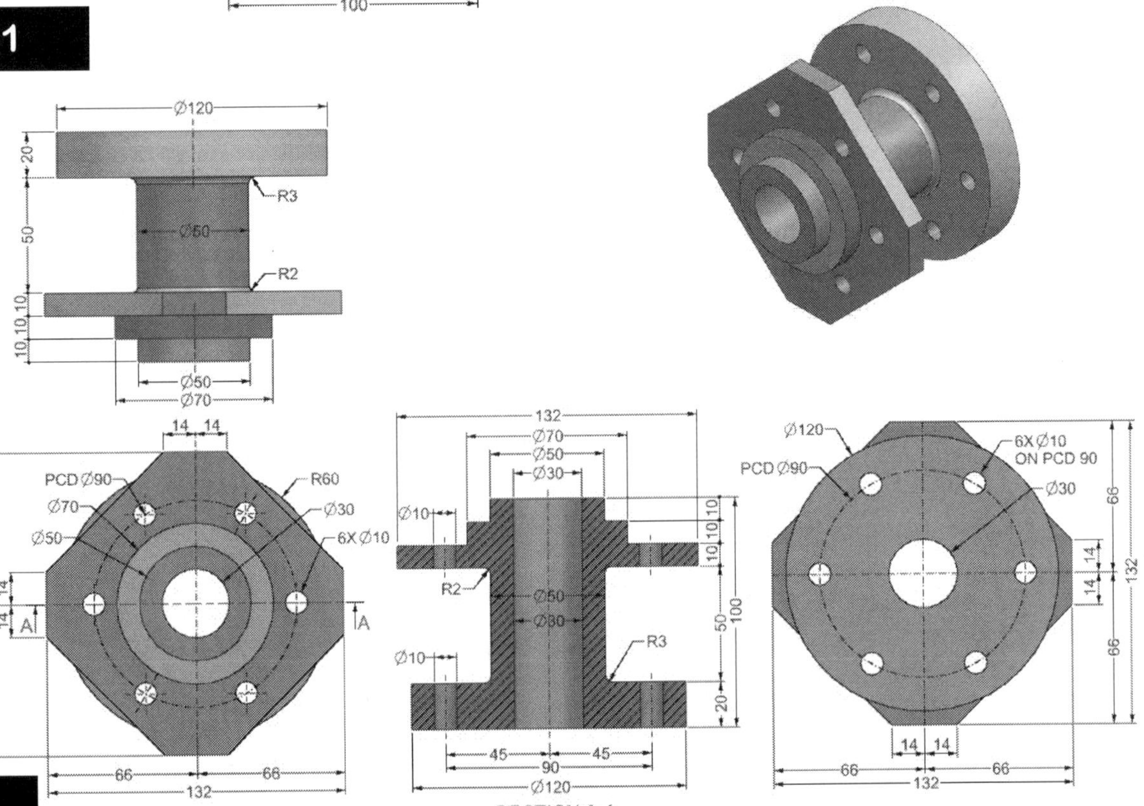
Ø120
20
50
R3
Ø50
R2
10 10 10
Ø50
Ø70
14 14
PCD Ø90
R60
Ø70
Ø30
Ø50
6X Ø10
66
132
14
14
A
A
66
66
132
132
Ø70
Ø50
Ø30
Ø10
10 10 10
R2
Ø50
Ø80
R3
Ø10
45
90
45
100
20
Ø120
SECTION A-A
Ø120
6X Ø10
ON PCD 90
PCD Ø90
Ø30
66
132
14
14
14
66
14 14
66
132
P-32

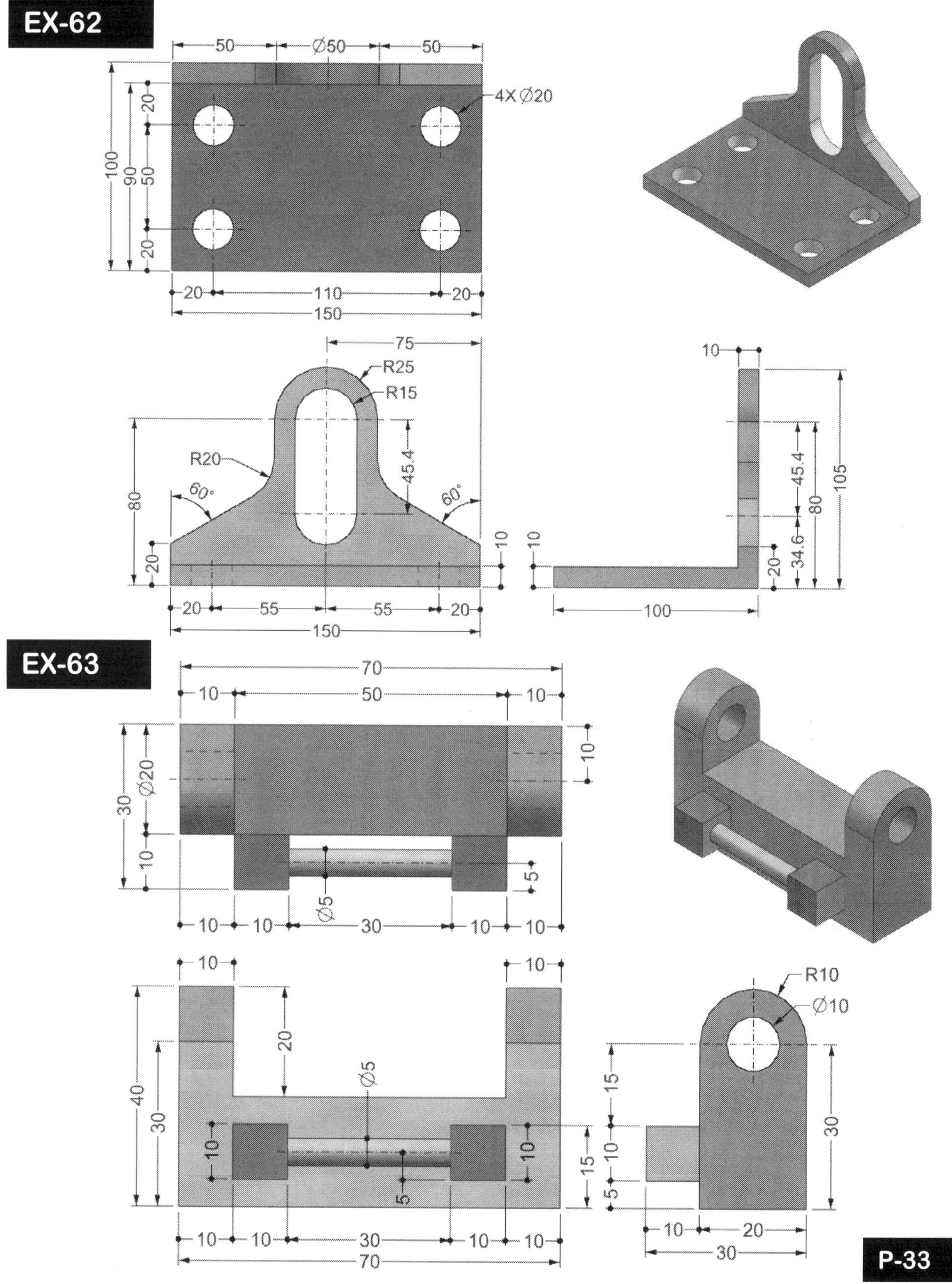

EX-62
50
Ø50
50
4X Ø20
20
100
90
50
20
20
110
20
150
75
R25
R15
R20
60°
80
60°
45.4
10
20
20
55
55
20
150
10
10
105
45.4
80
20
34.6
100
EX-63
70
10
50
10
10
30
Ø20
10
5
Ø5
10
10
30
10
10
10
10
20
40
30
10
Ø5
10
5
10
10
30
10
10
70
R10
Ø10
15
30
10
15
5
10
20
30
P-33

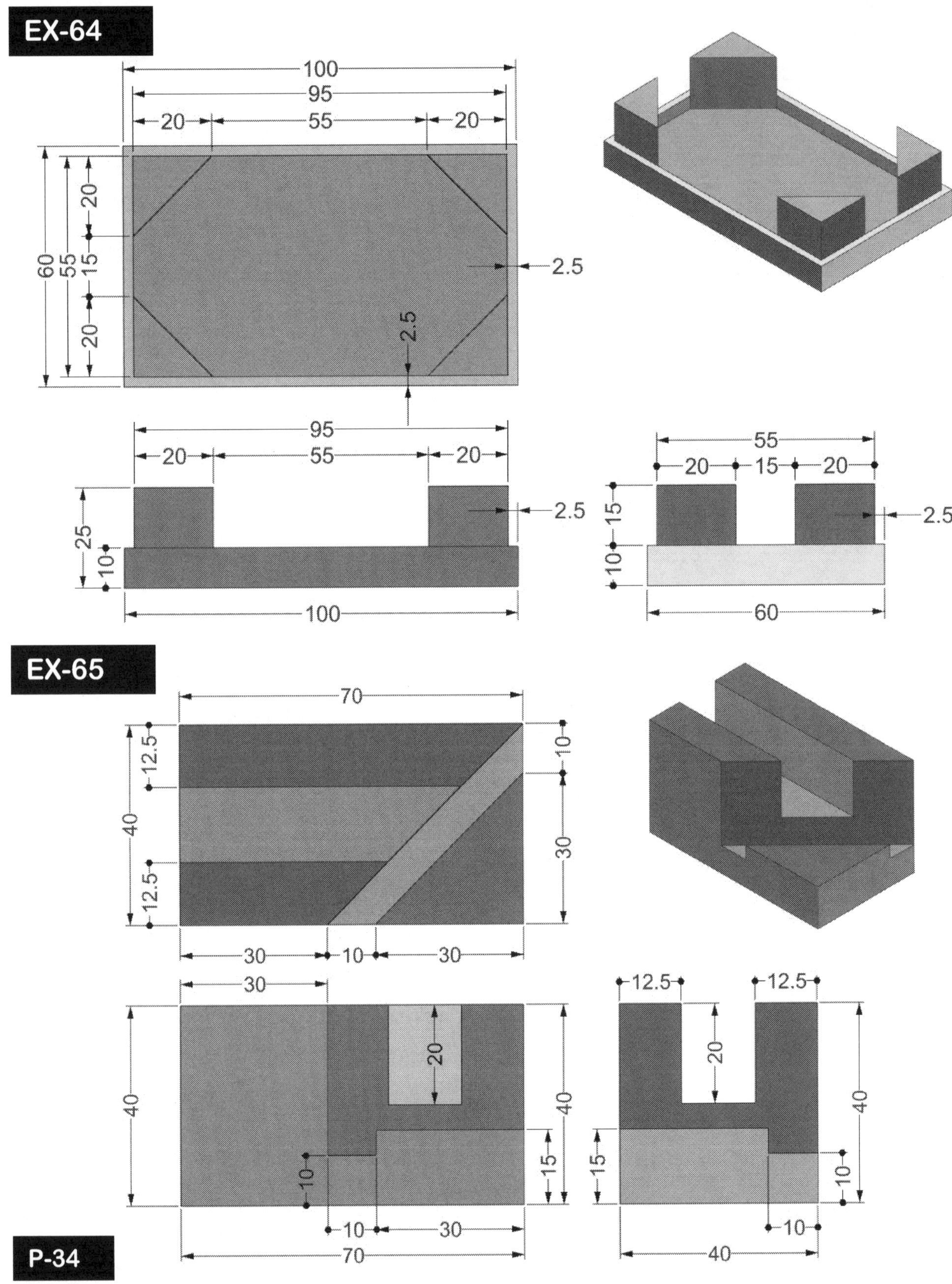
EX-64
100
95
20
55
20
60
55
20
15
20
2.5
2.5
95
20
55
20
25
10
2.5
100
55
20
15
20
15
10
2.5
60
EX-65
70
12.5
10
40
30
12.5
30
10
30
30
12.5
12.5
40
30
20
20
40
15
15
10
10
15
10
30
40
10
70
P-34

EX-66

30
R20
R20
R10
40
20
30
10
10
50
R10
10
R20
20
50
40
10
30
10
20
40

EX-67

70
45
R20
10
15
Ø20
20
40
10
45
65
45
Ø20
R25
10
40
30
20
10
Ø40
10
25
30
50
10
20

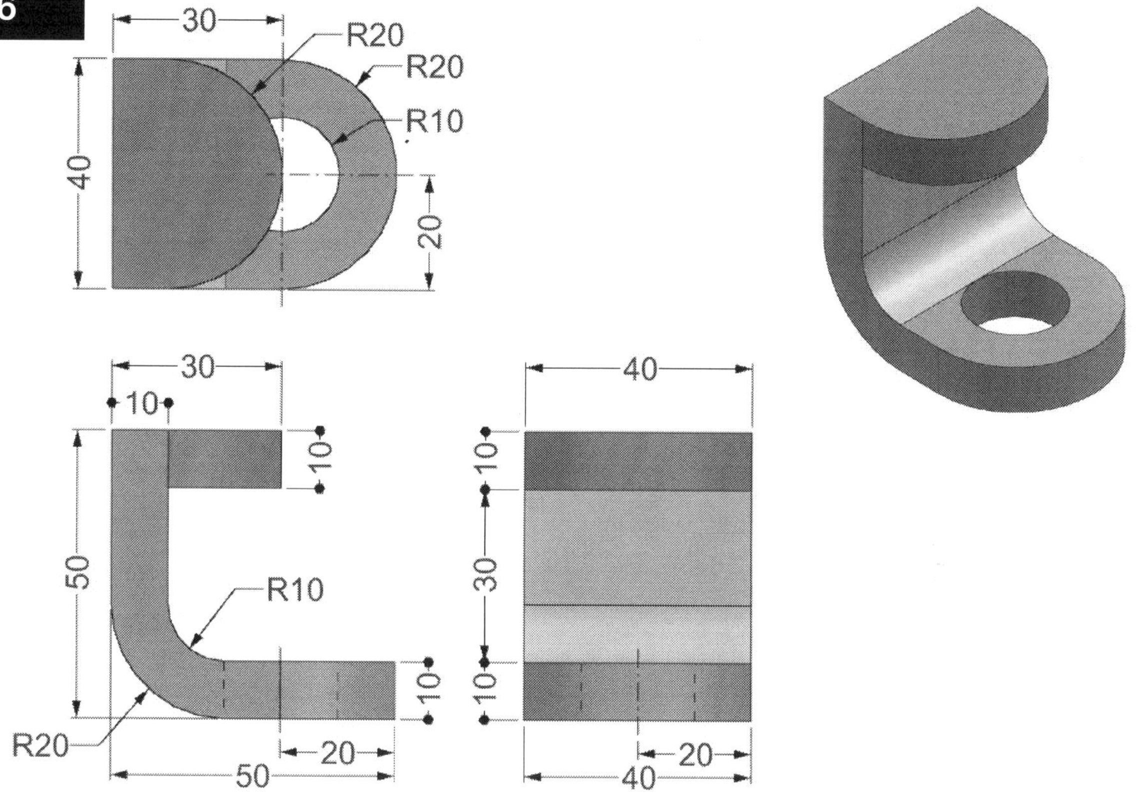

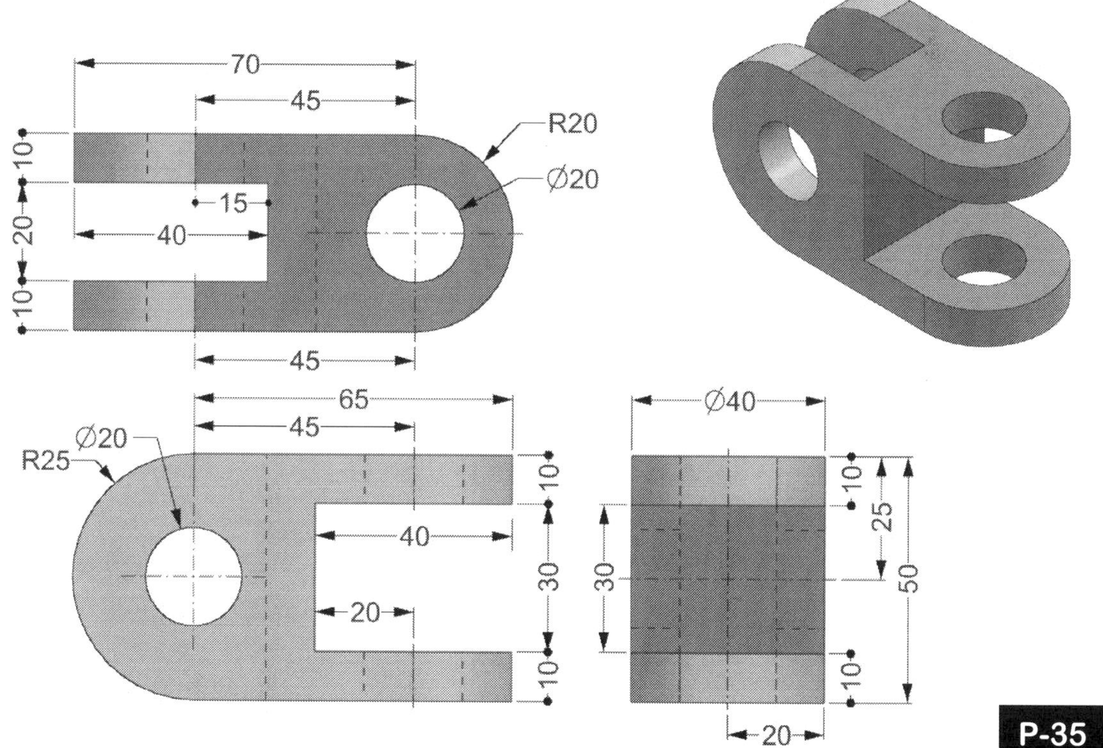

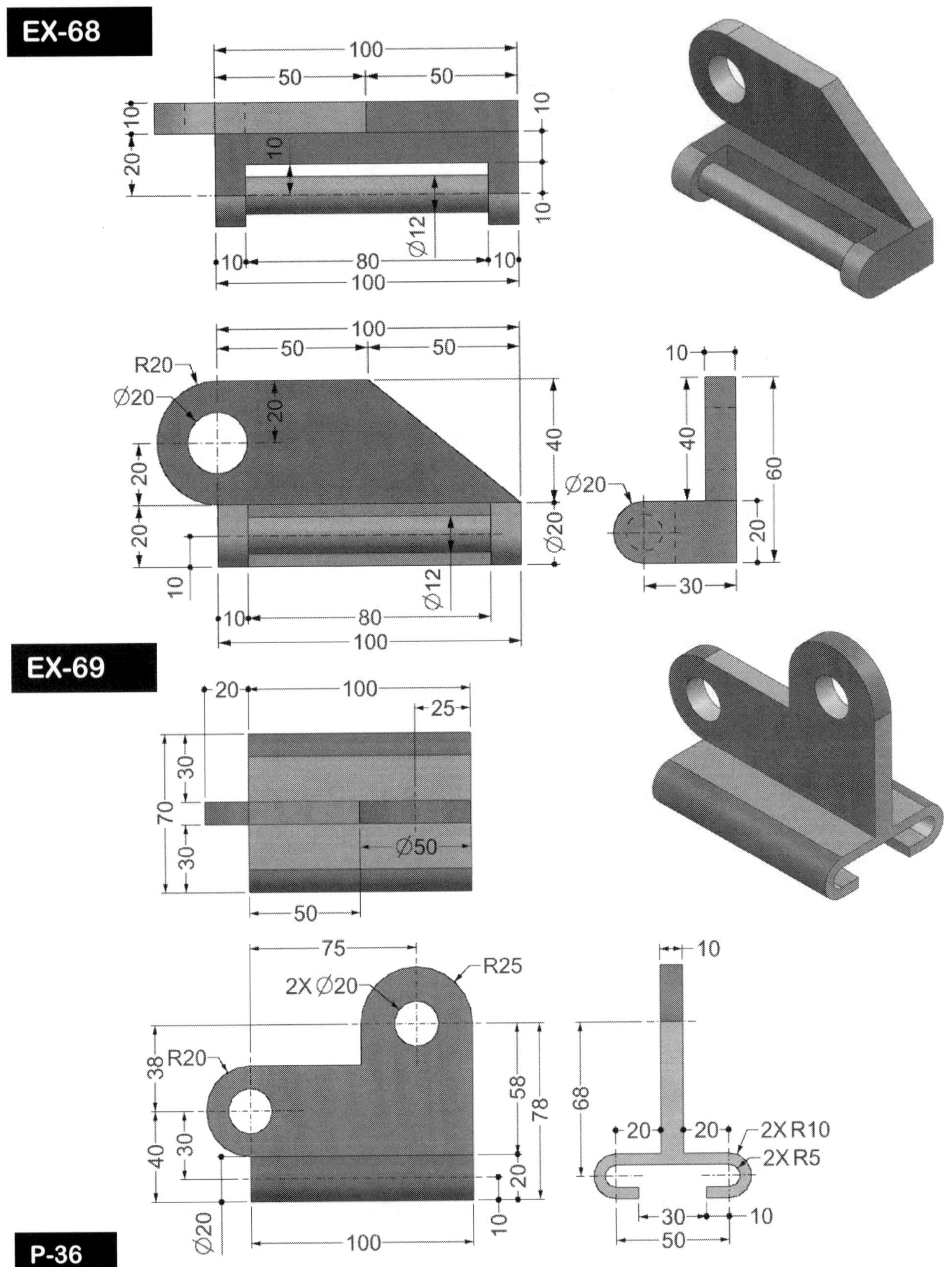

EX-68
100
50
50
10
20
10
10
10
10
80
Ø12
10
100
100
50
50
R20
Ø20
20
20
40
20
Ø20
Ø20
10
40
60
20
30
10
80
Ø12
10
100
EX-69
20
100
25
30
70
30
Ø50
50
75
R25
2X Ø20
R20
38
58
78
68
30
20
20
2X R10
2X R5
40
10
Ø20
100
30
10
50
P-36

EX-70
R10
Ø10
50
20
20
10
10
10
10
30
40
30
40
40
10
10
20
20
10
10
Ø10
50
20
10
20
10
20
20
10
10
10
10
40
60
30
10
R10
Ø10
30
10
30
50
10
20
30
10
20
10
10
10
20
Ø20
50
EX-71
200
30
70
70
30
4X Ø20
25
50
120
25
10
Ø70
Ø50
Ø70
Ø50
Ø70
R10
R10
R10
45
20
10
90
180
200
120
100
10
35
Ø70
45
20
10
100
10
120
P-37

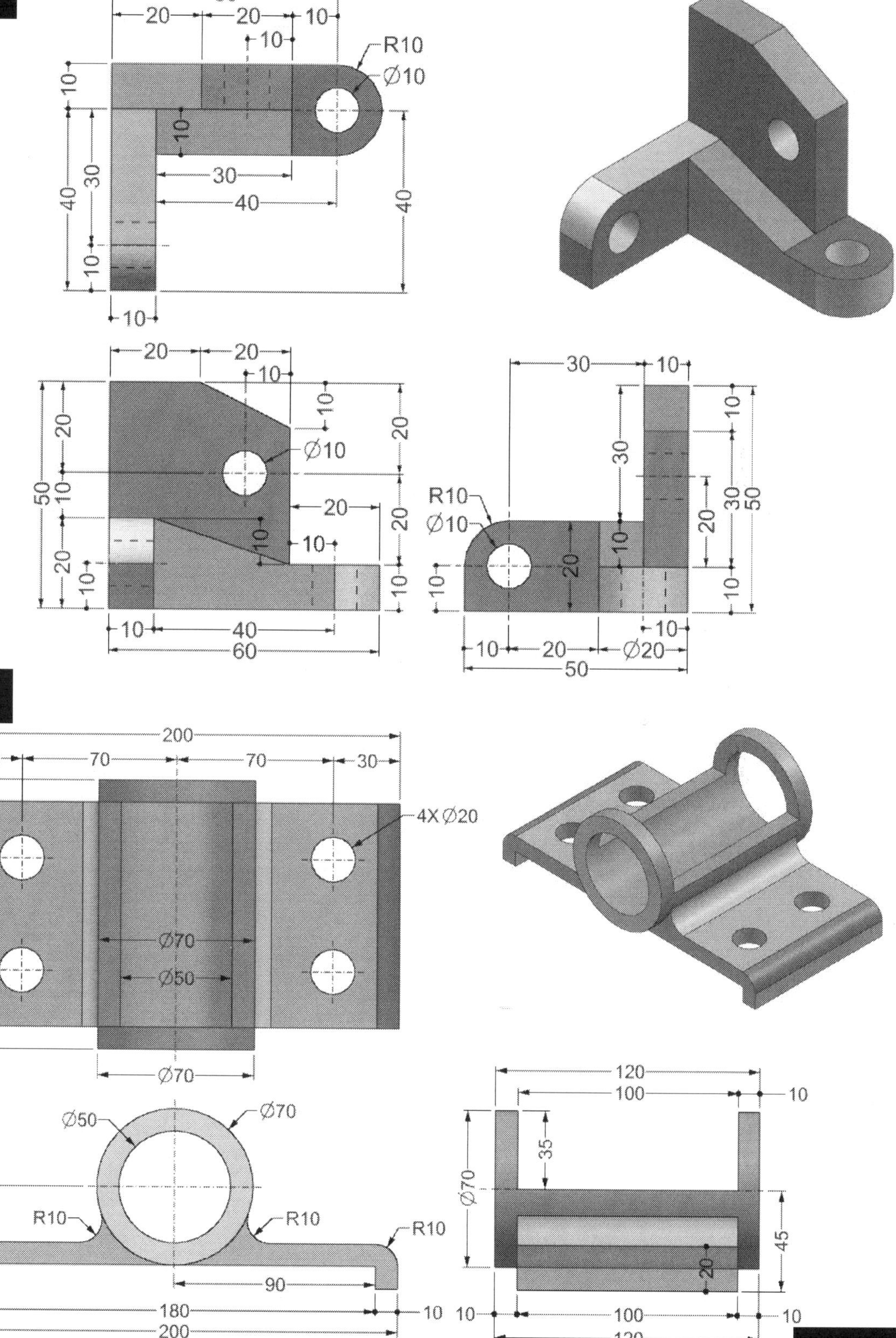

EX-72

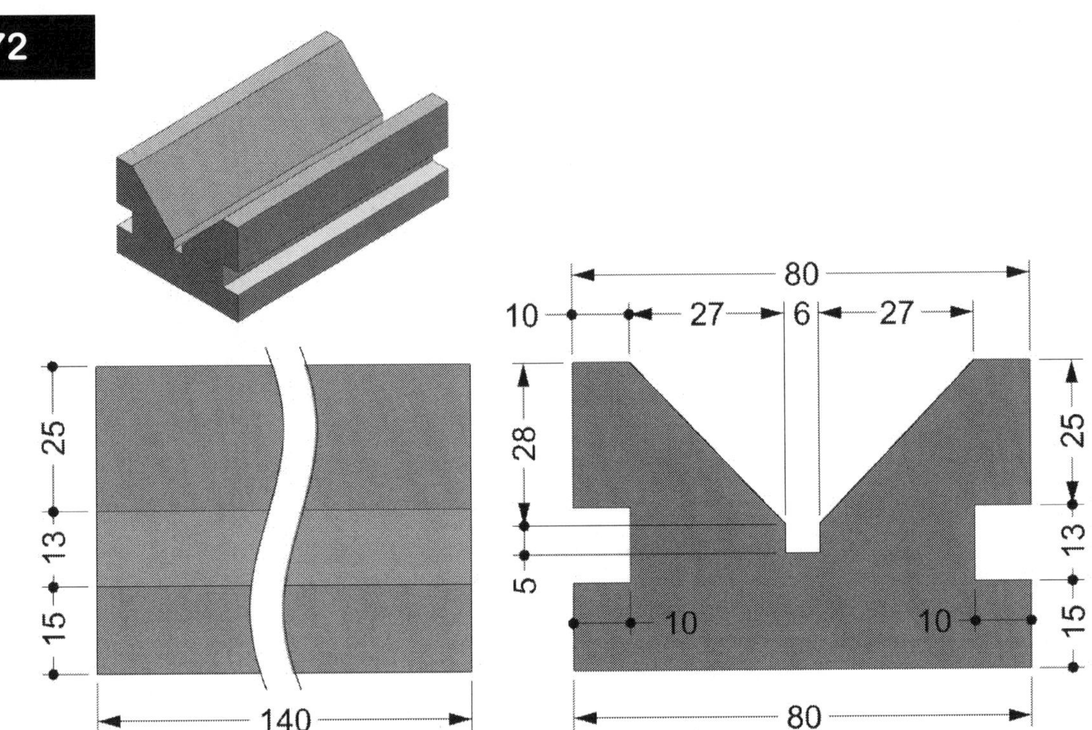

80
10
27
6
27
28
5
25
13
15
10
10
80
25
13
15
140

EX-73
15 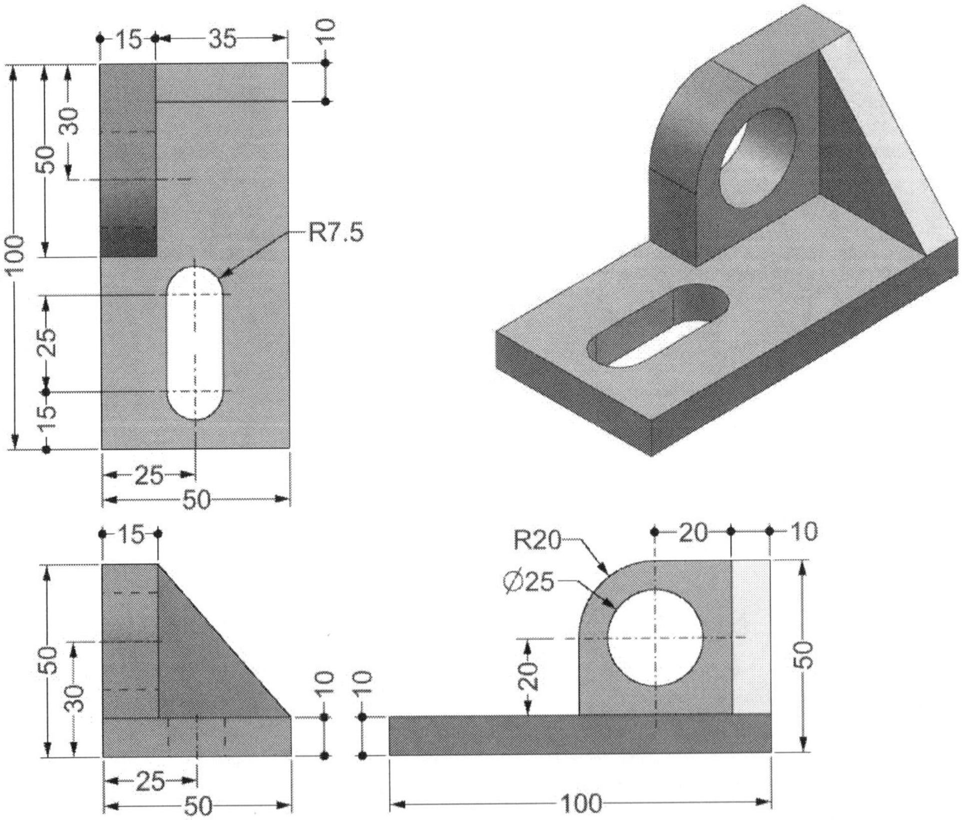
35
10
30
50
100
R7.5
25
15
25
50
15
50
30
10
10
R20
Ø25
20
10
20
50
25
50
100

P-38

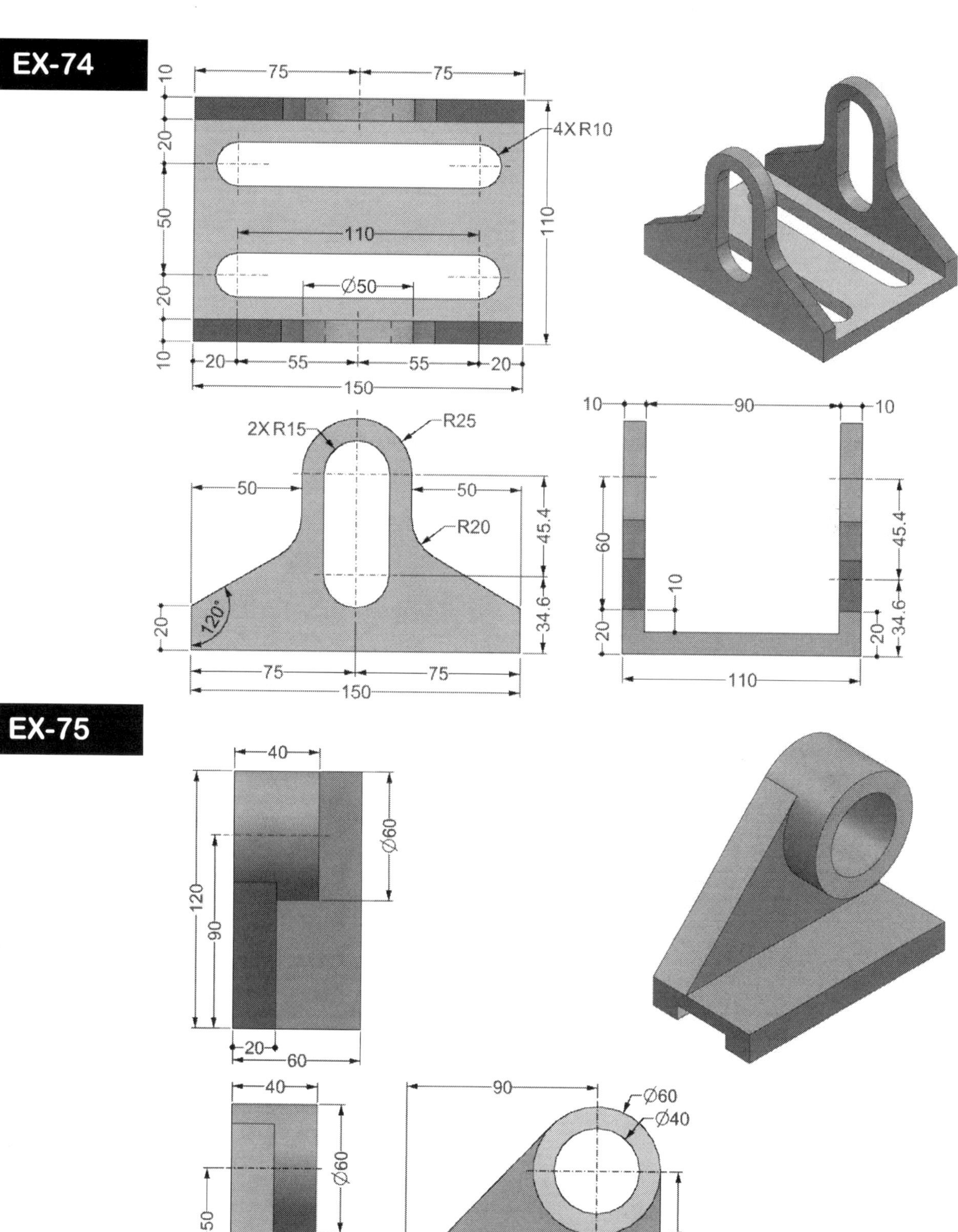
75
75
10
20
50
20
10
4X R10
110
110
Ø50
20
55
55
20
150
2X R15
R25
R20
50
50
45.4
34.6
20
120°
75
75
150
10
90
10
60
10
45.4
20
20
34.6
110
40
Ø60
120
90
20
60
40
Ø60
50
20
7.5
15
15
30
60
90
Ø60
Ø40
15
65
120

EX-76 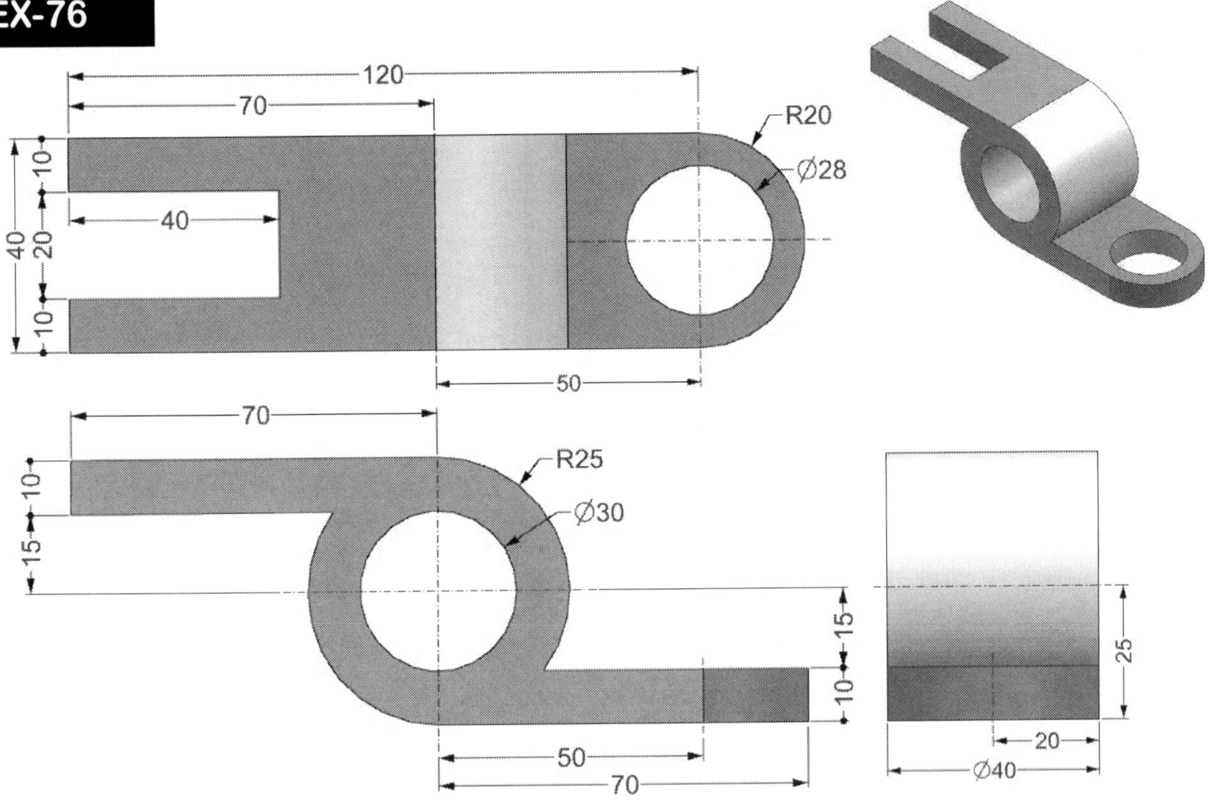
120
70
R20
10
40
20
Ø28
40
10
50
70
R25
10
Ø30
15
15
10
25
50
20
70
Ø40
EX-77 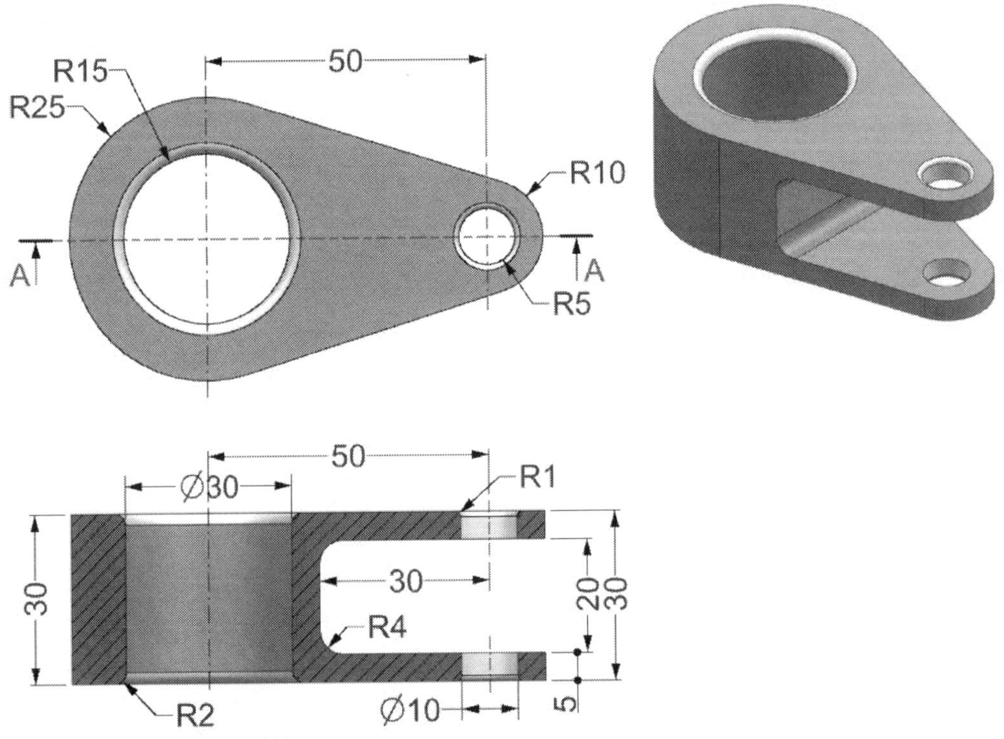
R15
50
R25
R10
A
A
R5
50
Ø30
R1
30
30
30
20
R4
R2
Ø10
5
SECTION A-A
P-40

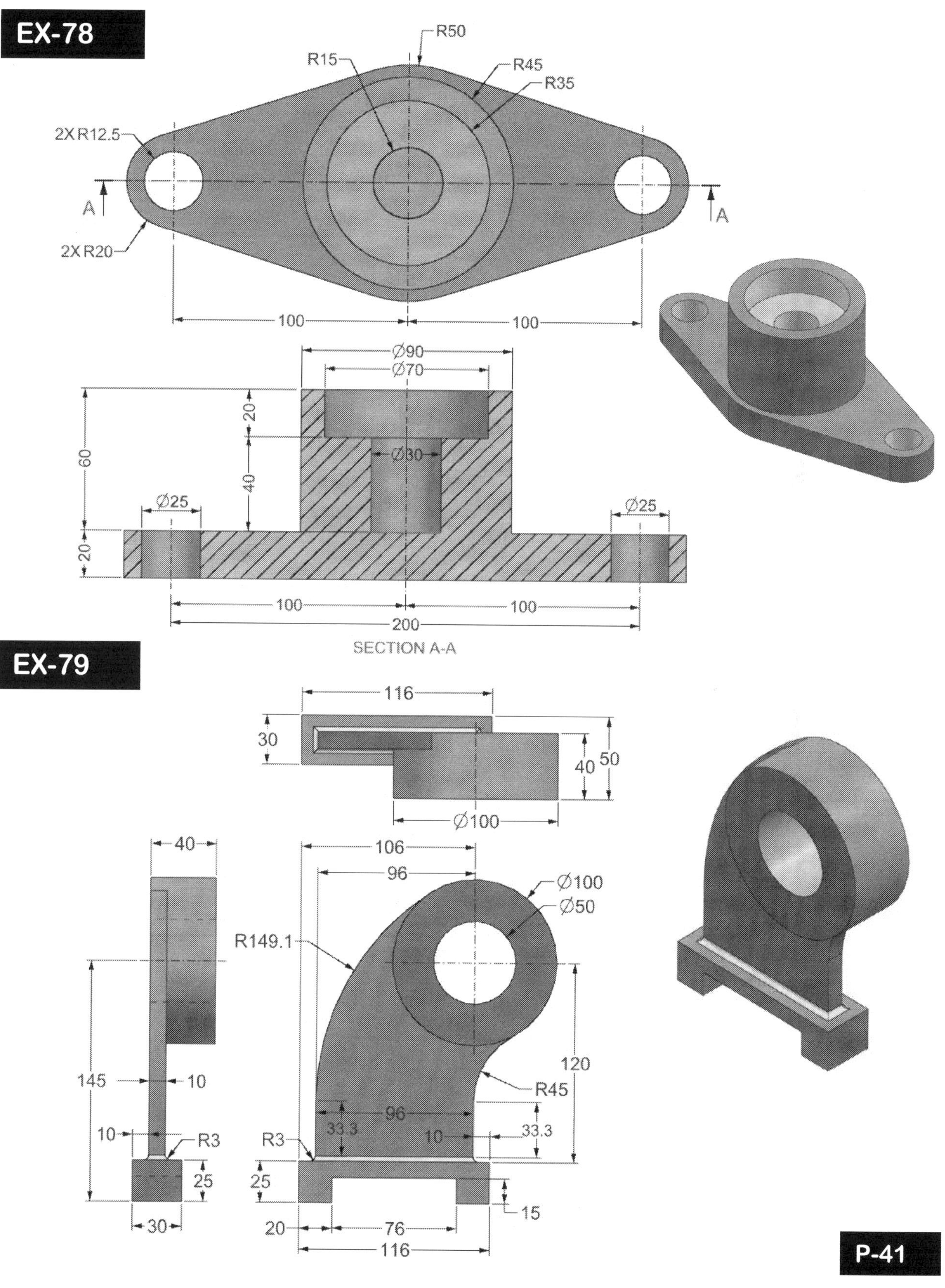

EX-78
R50
R15
R45
R35
2X R12.5
2X R20
A
A
100
100
Ø90
Ø70
20
Ø30
40
60
Ø25
Ø25
20
100
100
200
SECTION A-A
EX-79
116
30
40
50
Ø100
40
106
96
Ø100
Ø50
R149.1
145
10
120
10
R45
R3
96
33.3
33.3
R3
10
25
25
15
30
20
76
116
P-41

EX-80
6 HOLES, Ø10
ON DIA 32 PCD
4 HOLES, Ø8.6
ON DIA 54 PCD
Ø70
Ø16
A
A
Ø54
Ø32
CHAMFER 0.5 X 45°
4X Ø8.6
Ø16
6X Ø10
10
5
5
SECTION A-A
(SCALE 1:1)

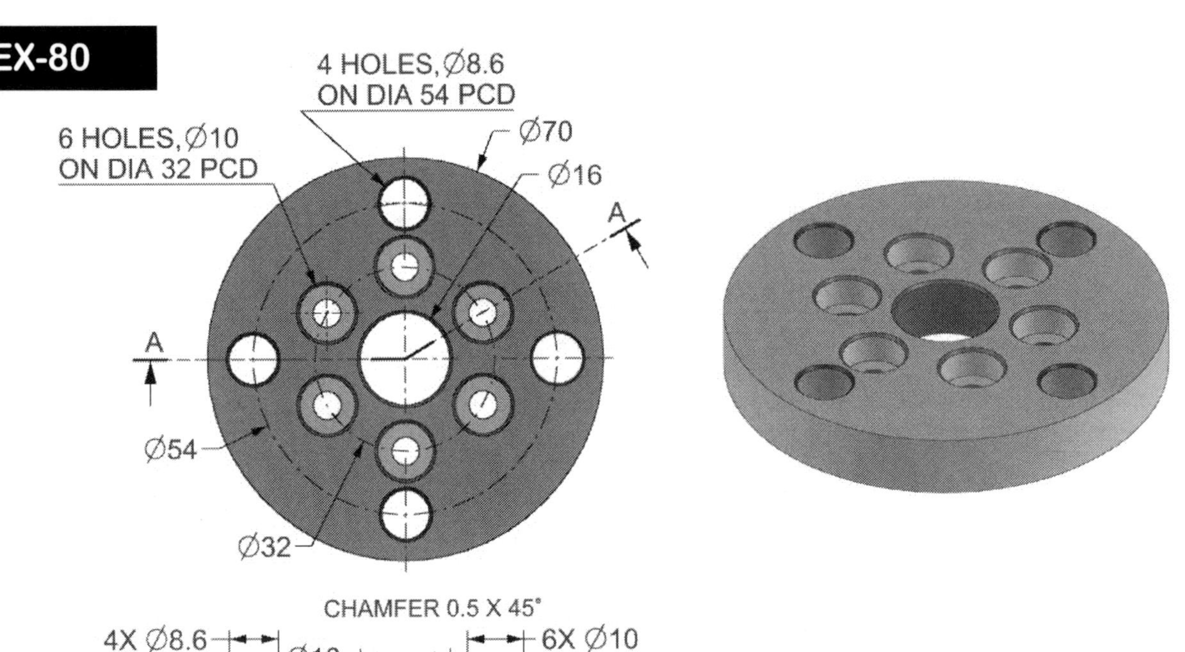

EX-81 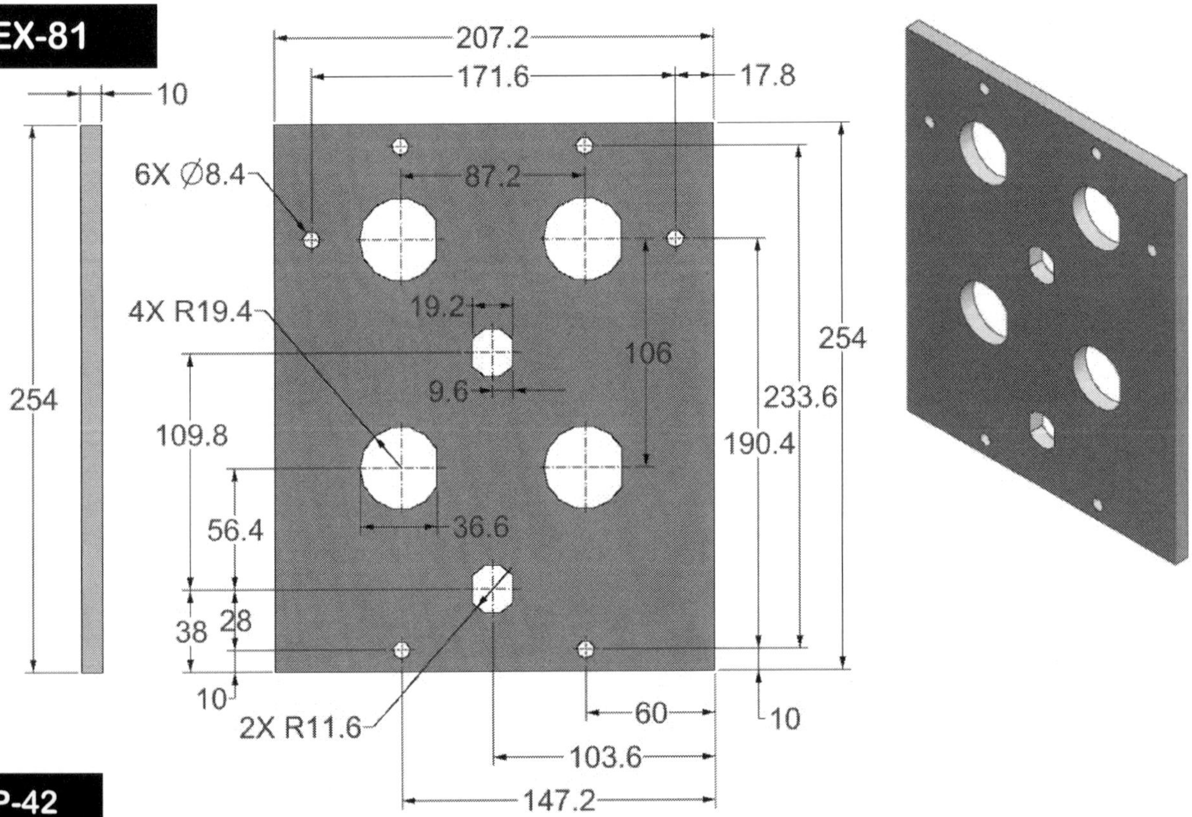
10
6X Ø8.4
4X R19.4
254
109.8
56.4
38
28
10
2X R11.6
207.2
171.6
17.8
87.2
19.2
9.6
106
233.6
190.4
254
36.6
60
10
103.6
147.2
P-42

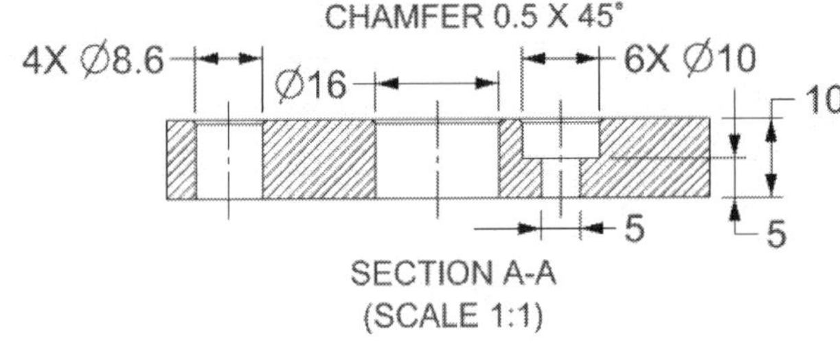

EX-82 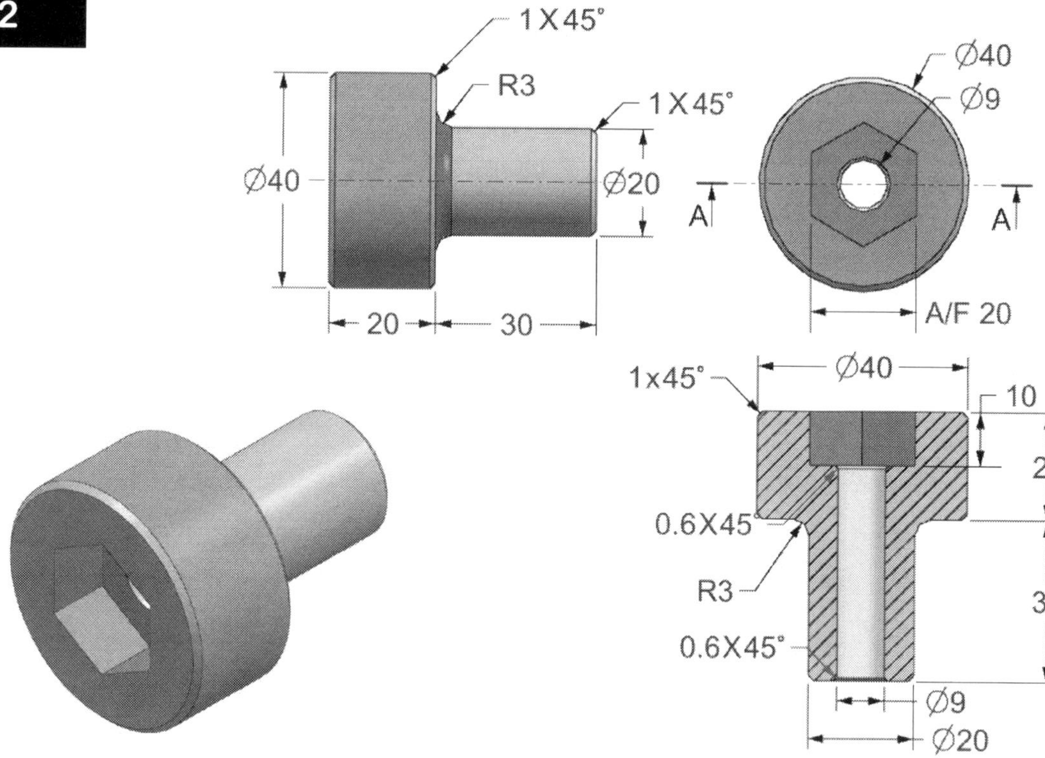
1 X 45°
R3
1 X 45°
Ø40
Ø20
20
30
Ø40
Ø9
A
A
A/F 20
1x45°
Ø40
10
20
0.6X45
R3
30
0.6X45°
Ø9
Ø20
SECTION A-A
(SCALE 1:1)

EX-83 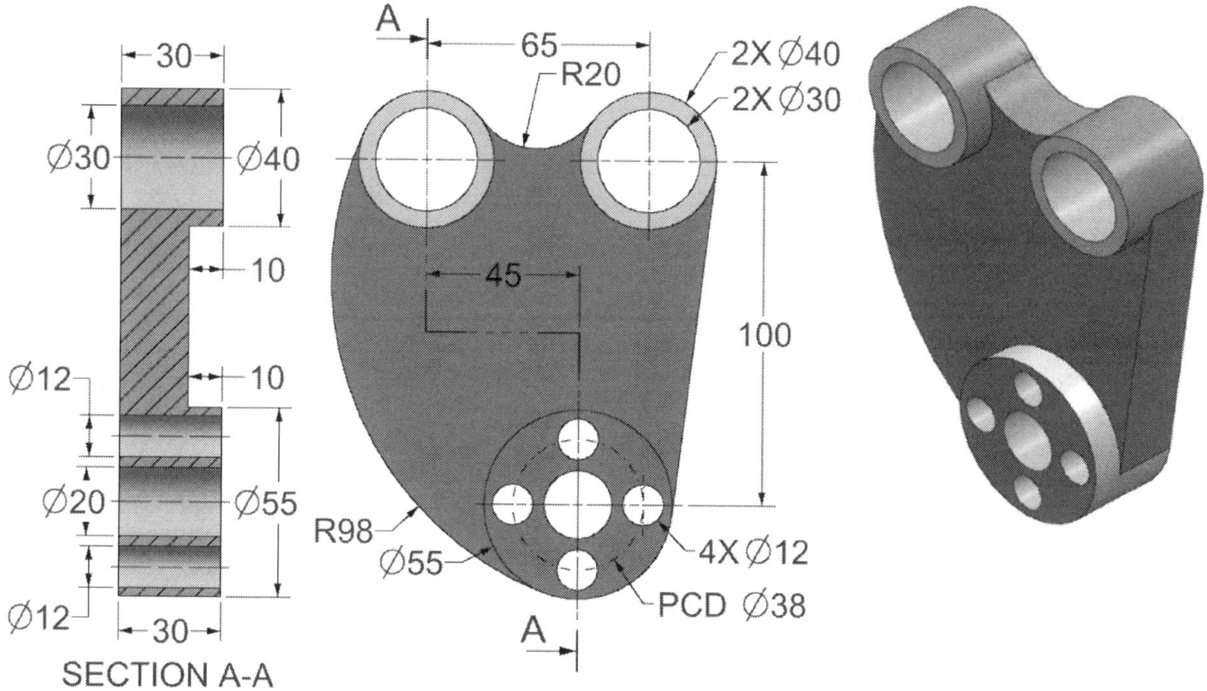
30
R20
A
65
2X Ø40
2X Ø30
Ø30
Ø40
10
45
10
100
Ø12
Ø20
Ø55
R98
Ø55
4X Ø12
Ø12
30
A
PCD Ø38
SECTION A-A
(SCALE 1:1)
P-43

EX-84

1
8
4X R0.7
12X Ø0.8
1.7
1
4.4
0.7
0.7
6.5
4.5
30.5
2.1
4X R0.5
Ø2.3
4
8
5
12.3
19.7
27.1
34.4
35.9

EX-85

30
30
4
4
8.5
8.5
40
4
40
20
A
A
8.5
8.5
40
4
4
8
4
60
SECTION A-A
(SCALE 1:1)

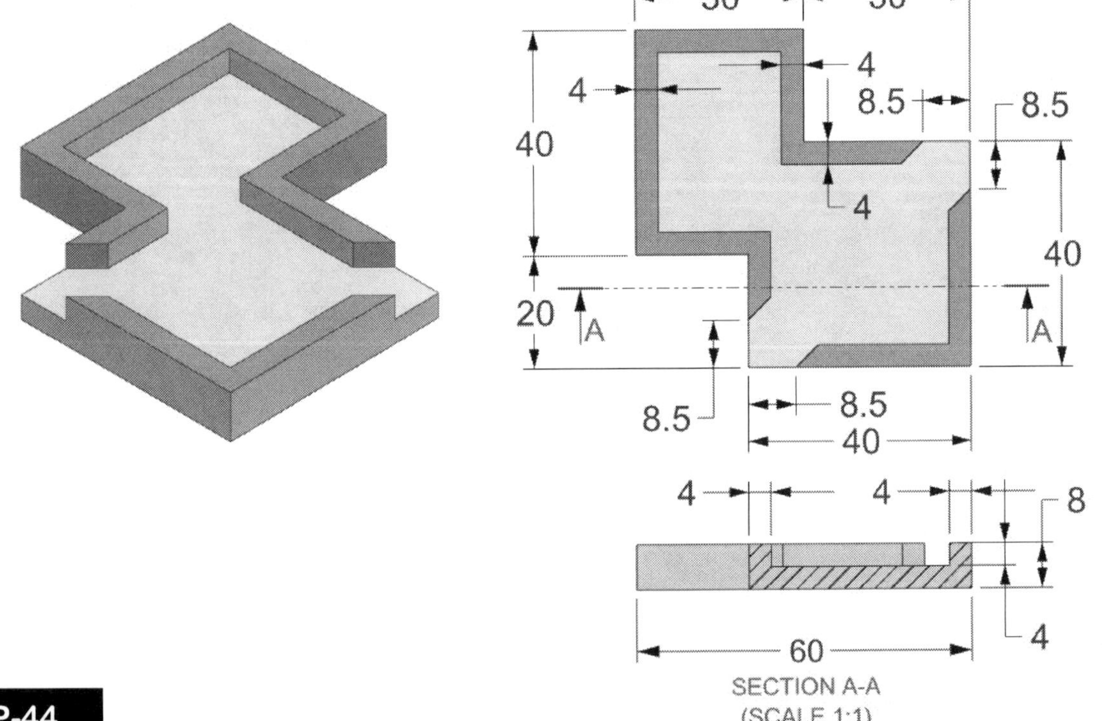

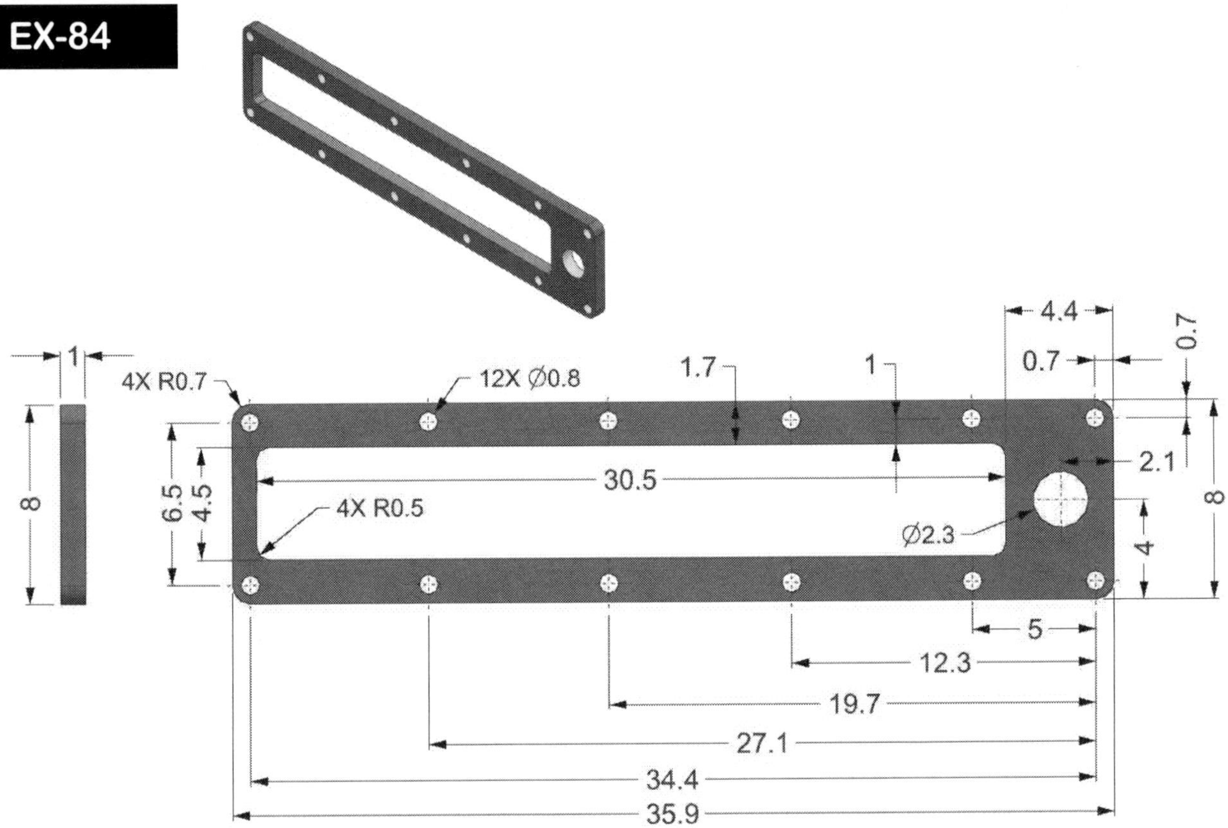

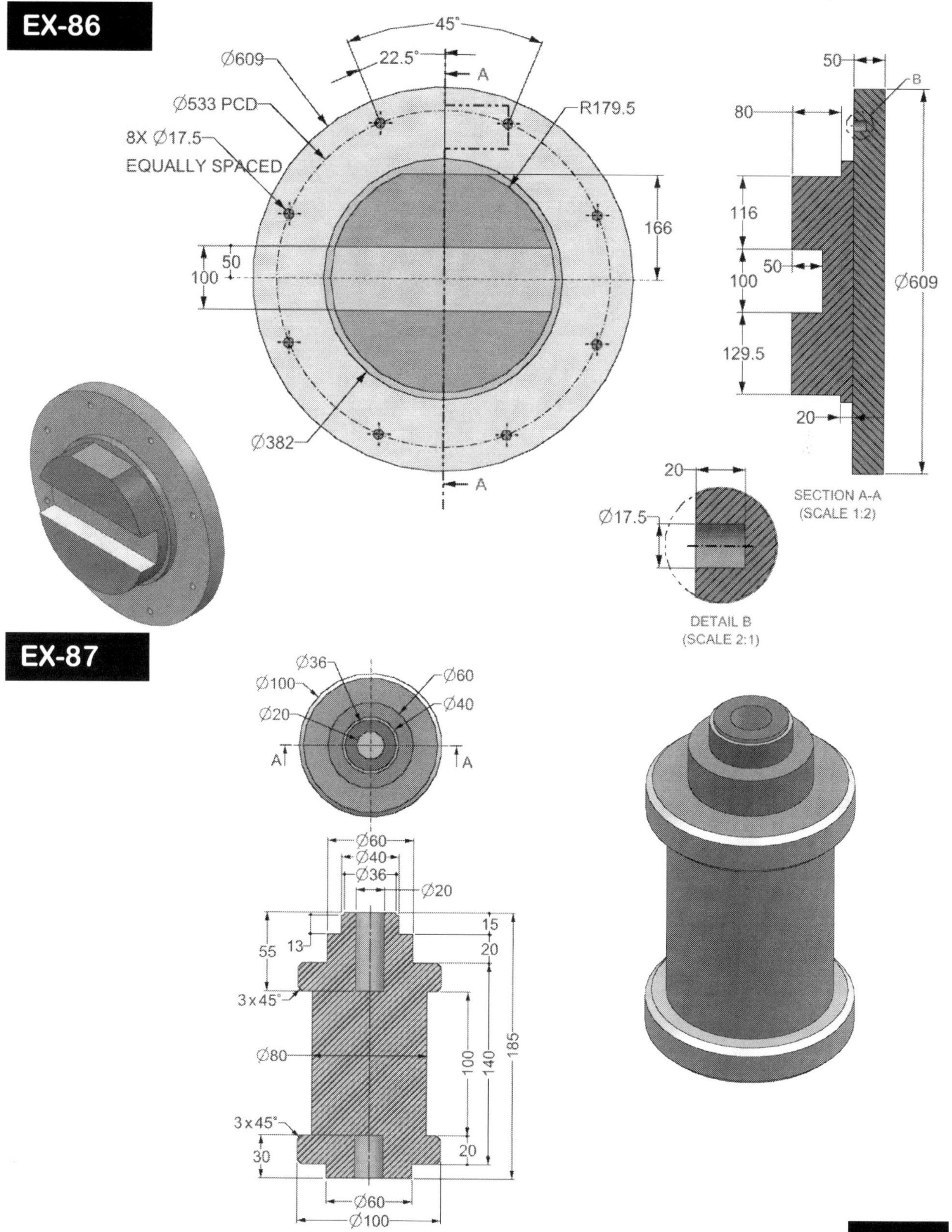

EX-86
Ø609
Ø533 PCD
8X Ø17.5
EQUALLY SPACED
45°
22.5°
A
R179.5
166
50
100
Ø382
A
50
80
116
50
100
129.5
20
B
Ø609
SECTION A-A
(SCALE 1:2)
20
Ø17.5
DETAIL B
(SCALE 2:1)
EX-87
Ø36
Ø100
Ø20
Ø60
Ø40
A
A
Ø60
Ø40
Ø36
Ø20
15
20
55
13
3 x 45°
Ø80
100
140
185
3 x 45°
30
20
Ø60
Ø100
SECTION A-A
P-45

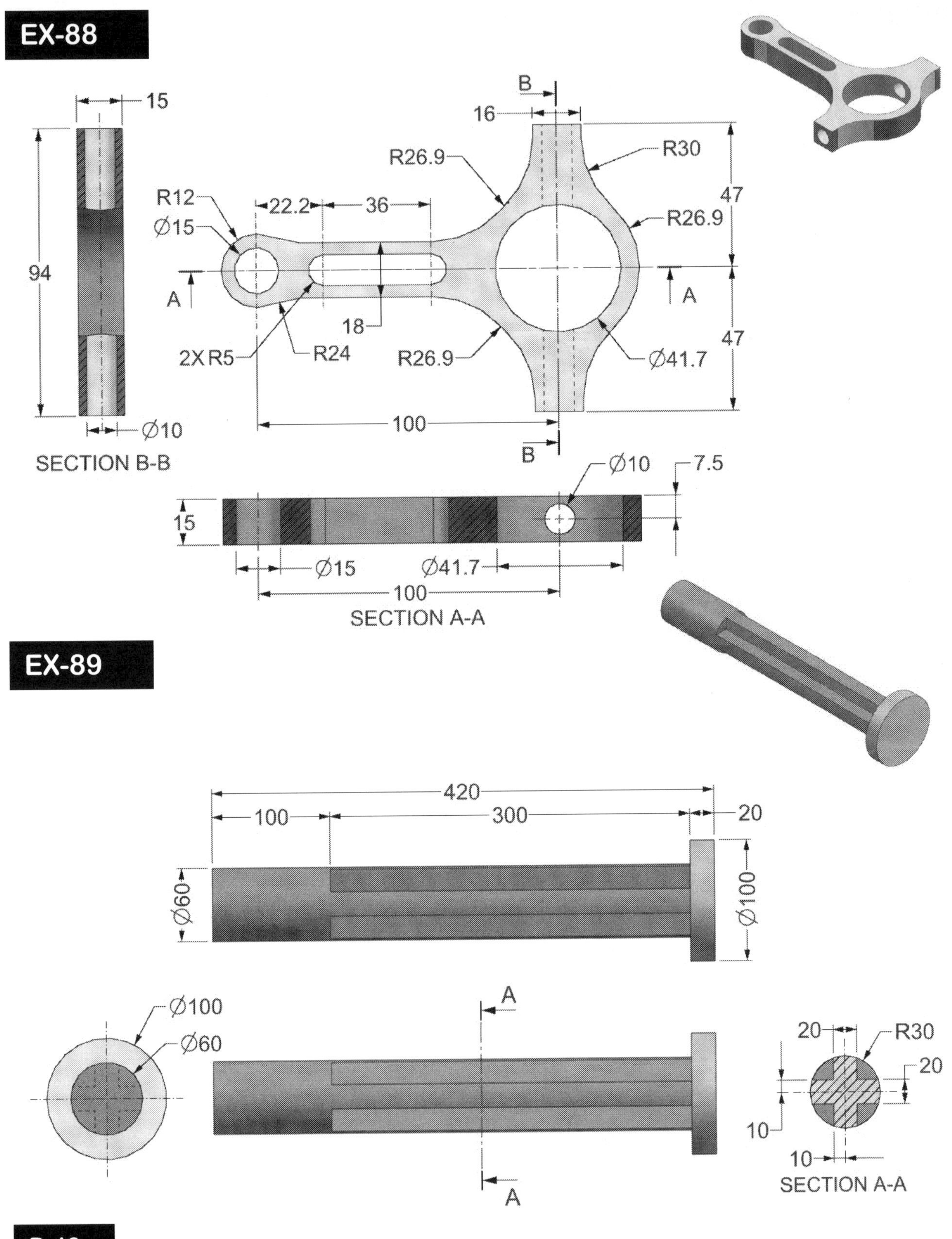

EX-88

15

R12
Ø15

R26.9

B

16

R30

R26.9

22.2

36

47

94

A

A

18

47

2X R5

R24

R26.9

Ø41.7

100

B

Ø10

SECTION B-B

Ø10

7.5

15

Ø15

Ø41.7

100

SECTION A-A

EX-89

420

100

300

20

Ø60

Ø100

A

Ø100

Ø60

A

20

R30

20

10

10

SECTION A-A

A

A

P-46

EX-90
2X Ø32 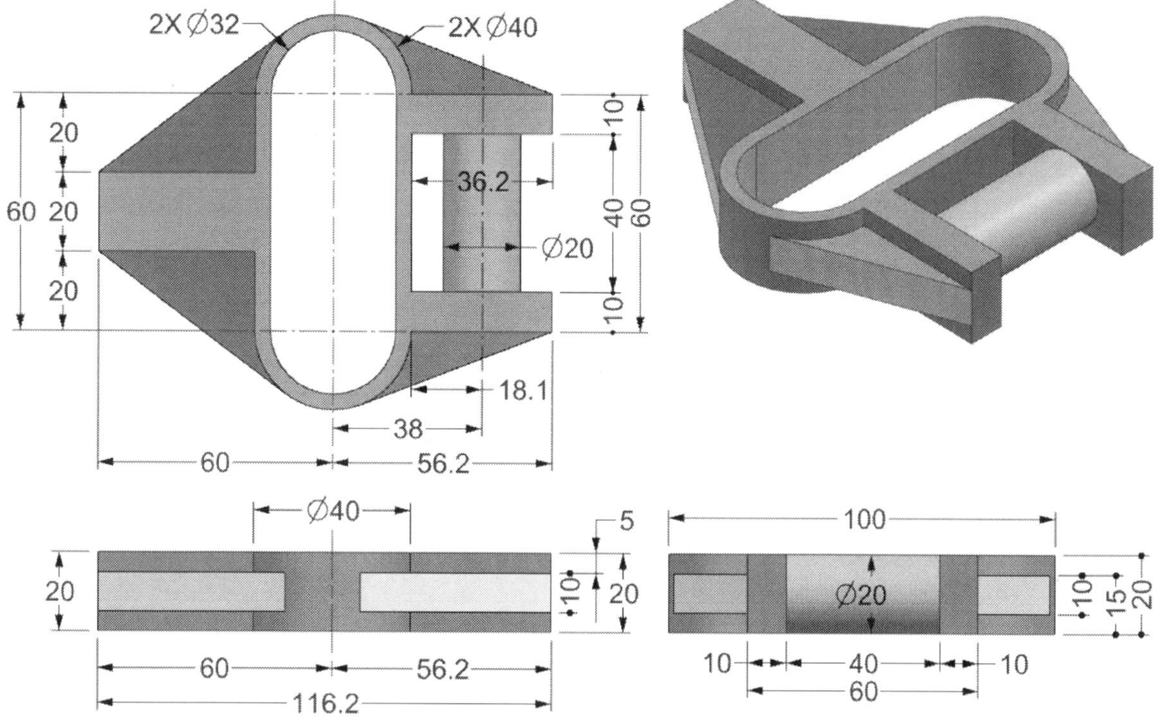
2X Ø40
20
20
60
20
10
36.2
Ø20
40
60
10
18.1
38
60
56.2
Ø40
20
5
10
20
60
56.2
116.2
100
Ø20
10
15
20
10
40
10
60

EX-91
30 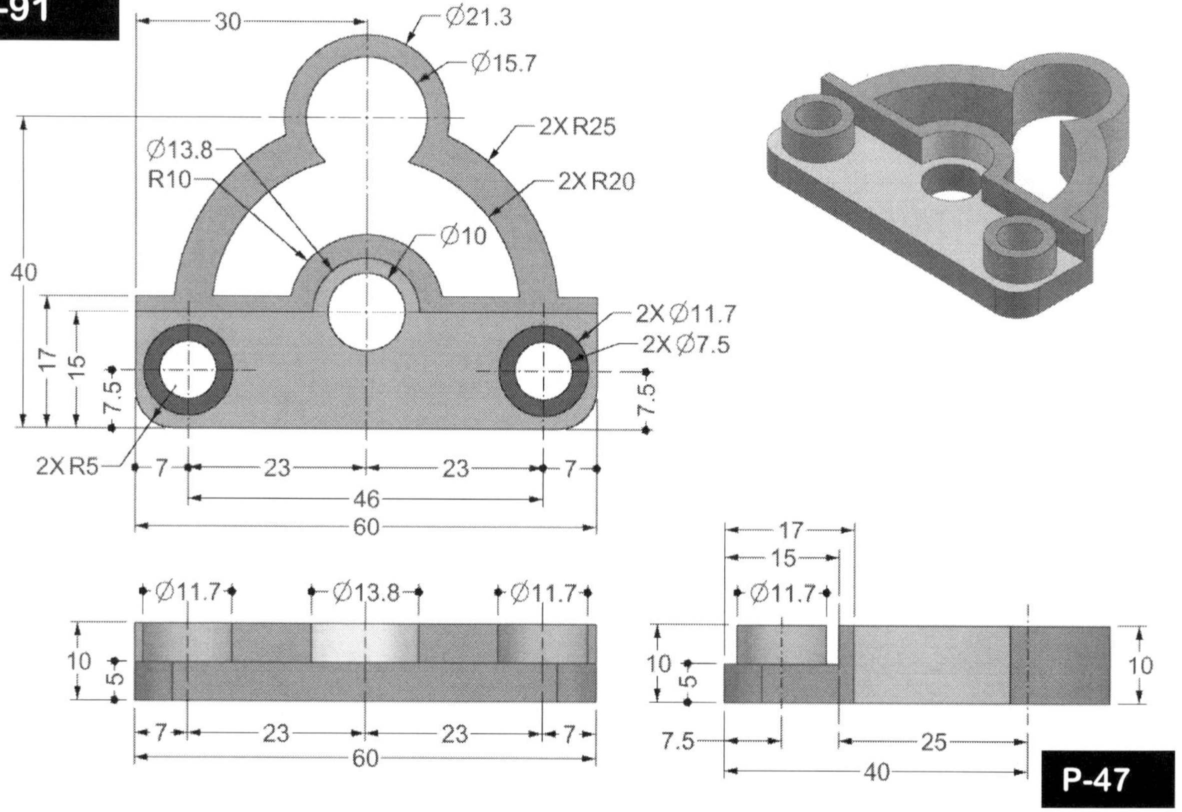
Ø21.3
Ø15.7
2X R25
Ø13.8
R10
2X R20
Ø10
40
2X Ø11.7
2X Ø7.5
17
15
7.5
7.5
2X R5
7
23
23
7
46
60
Ø11.7
Ø13.8
Ø11.7
10
5
7
23
23
7
60
17
15
Ø11.7
10
5
10
7.5
25
40

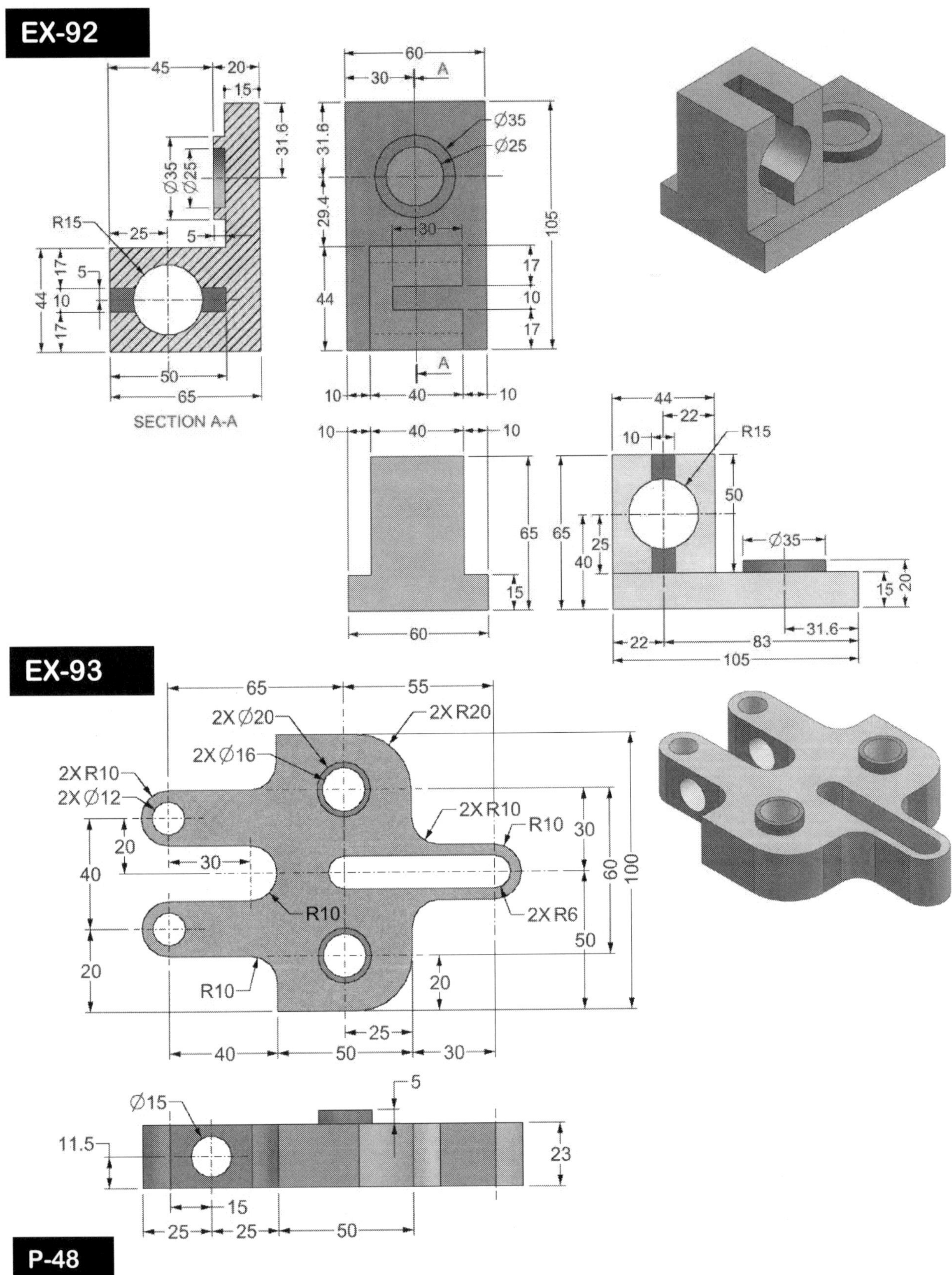

EX-92
45
20
15
31.6
Ø35
Ø25
R15
25
5
5
44
17
10
17
50
65
SECTION A-A
60
30
A
31.6
Ø35
Ø25
29.4
30
105
44
17
10
17
A
10
40
10
10
40
10
65
15
60
44
22
10
R15
50
Ø35
65
25
40
15
20
22
83
31.6
105
EX-93
65
55
2X Ø20
2X R20
2X Ø16
2X R10
2X Ø12
2X R10
R10
30
30
60
100
40
20
R10
2X R6
50
20
20
R10
25
40
50
30
Ø15
5
11.5
23
25
15
25
50
P-48

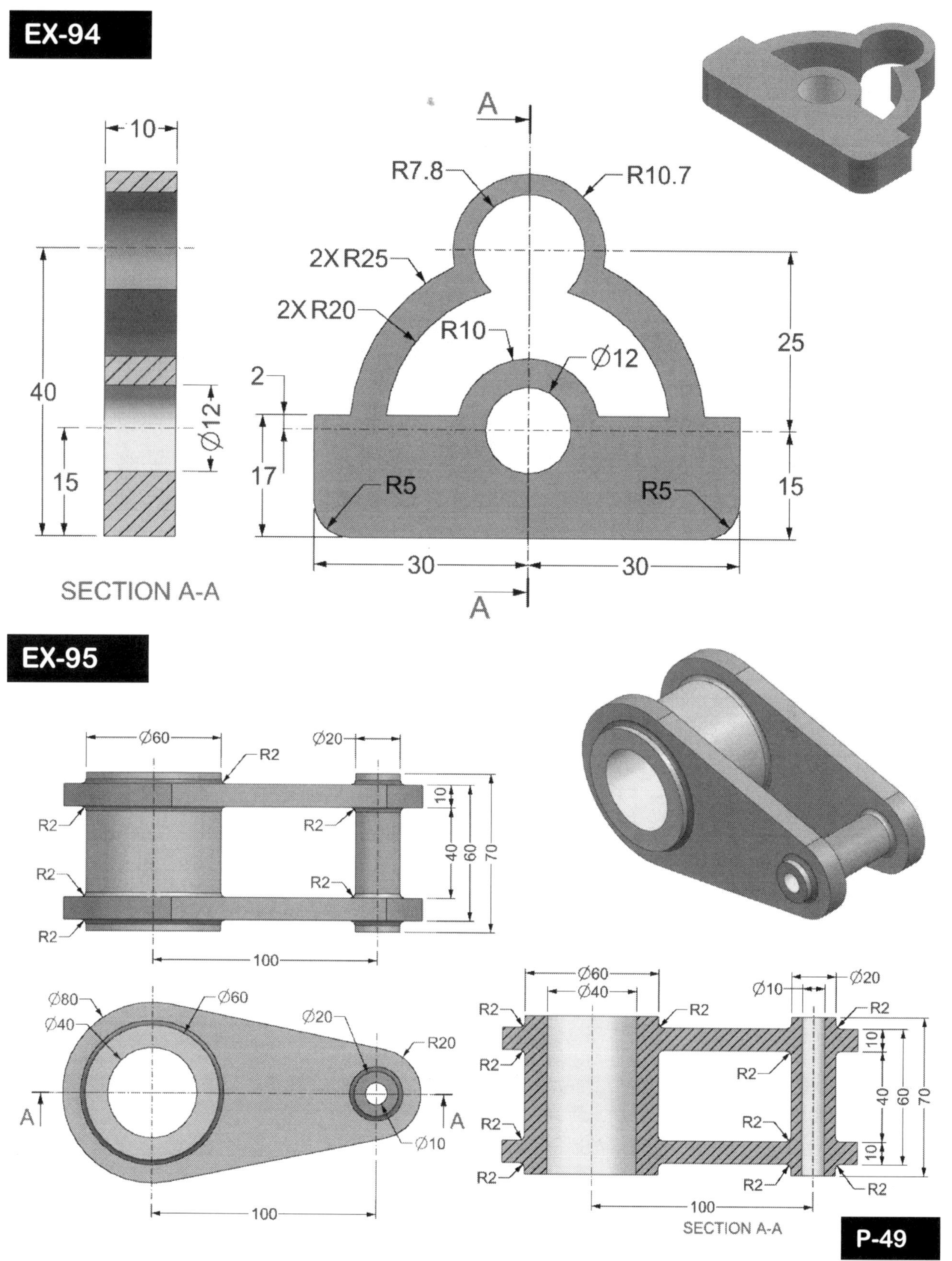
EX-94
10
40
15
Ø12
SECTION A-A
A
R7.8
R10.7
2X R25
2X R20
R10
Ø12
2
17
25
15
R5
R5
30
30
A
EX-95
Ø60
Ø20
R2
R2
R2
R2
R2
10
40
60
70
100
Ø80
Ø60
Ø40
Ø20
R20
A
A
Ø10
100
Ø60
Ø40
Ø10
Ø20
R2
R2
R2
R2
R2
R2
R2
R2
R2
10
40
60
70
10
100
SECTION A-A
P-49

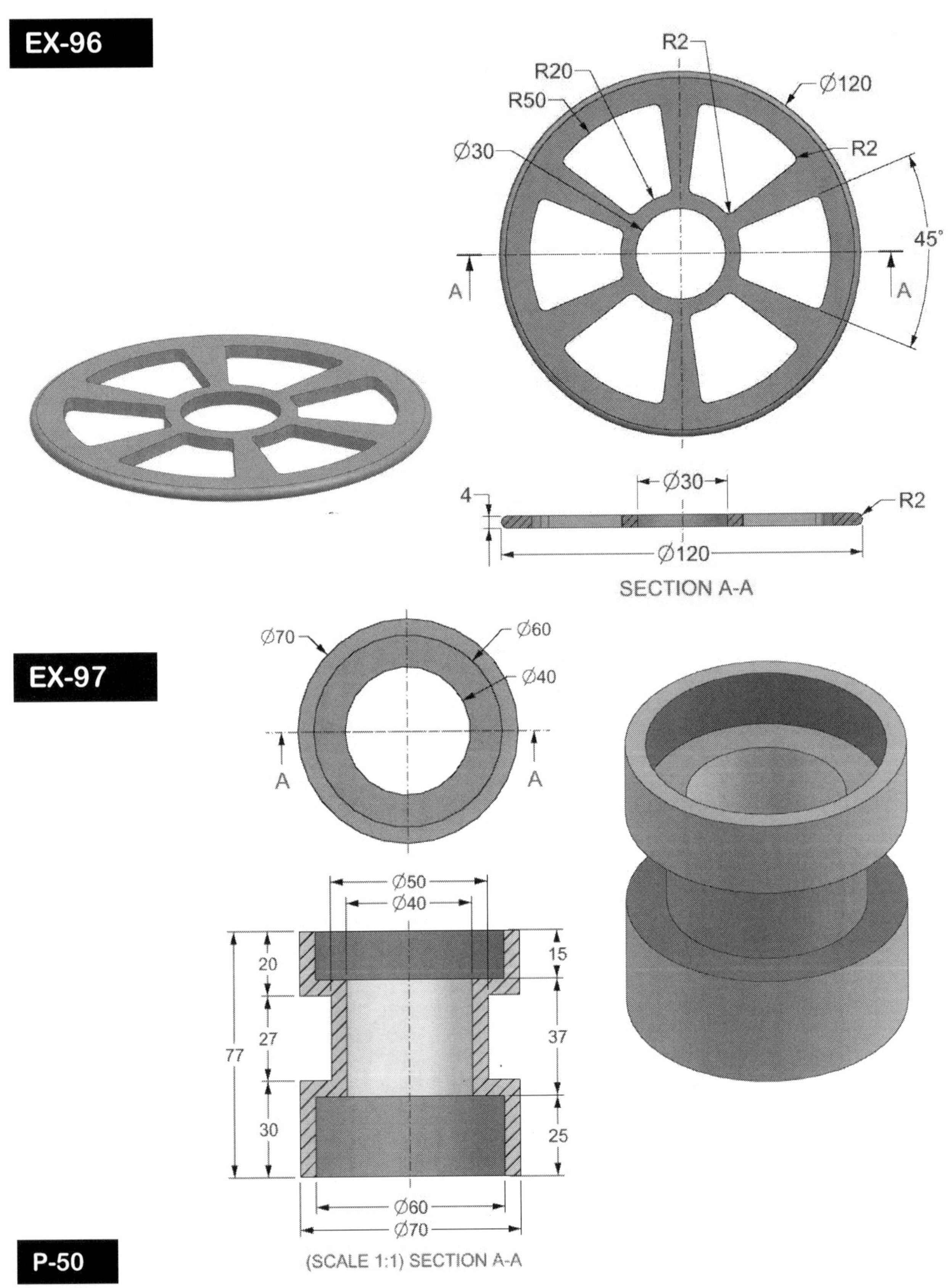

EX-96
R2
R20
R50
Ø120
Ø30
R2
45°
A
A
4
Ø30
R2
Ø120
SECTION A-A
EX-97
Ø70
Ø60
Ø40
A
A
Ø50
Ø40
20
15
27
37
77
30
25
Ø60
Ø70
(SCALE 1:1) SECTION A-A
P-50

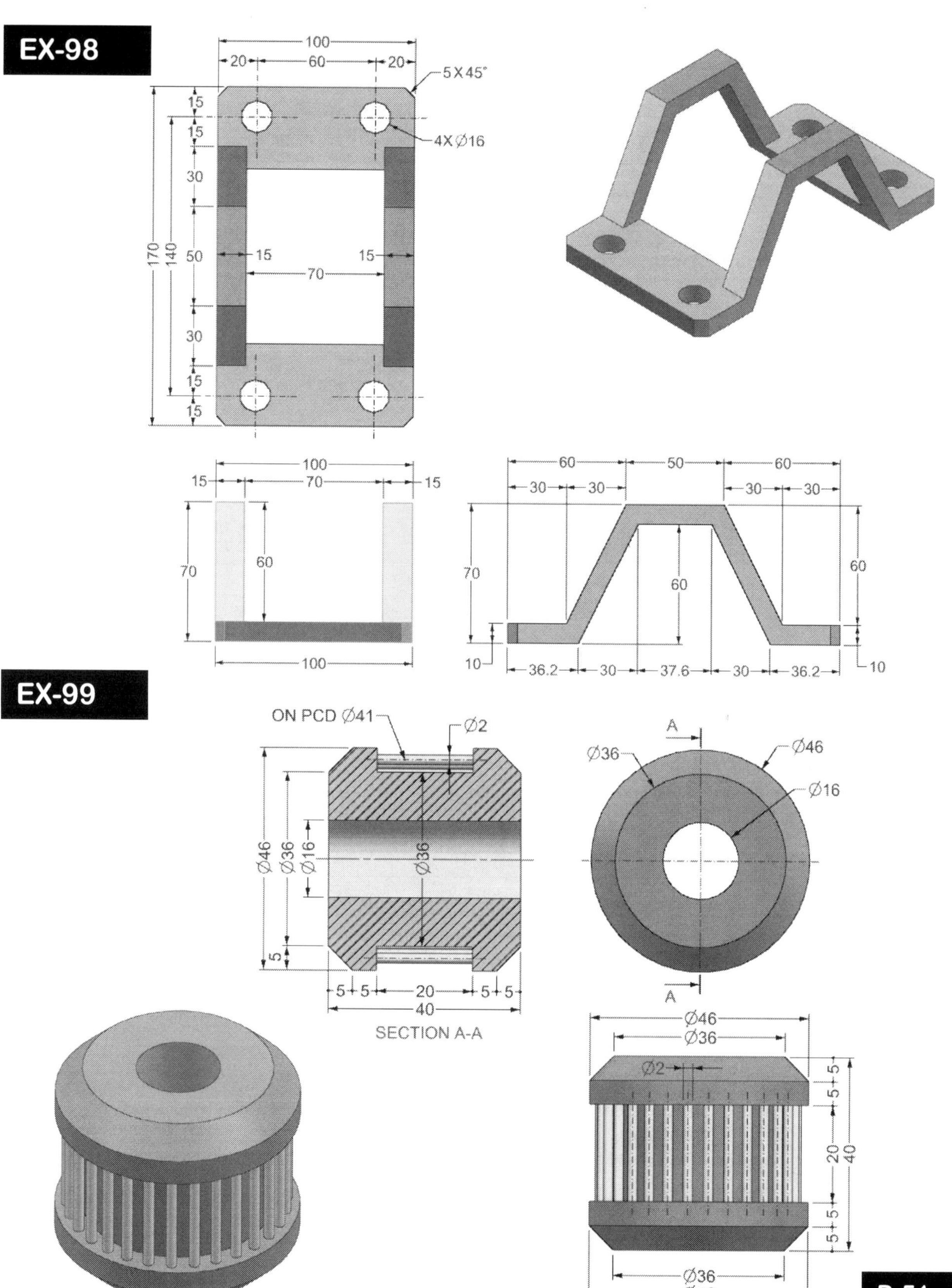

EX-98
100
20
60
20
5 X 45°
15
15
30
15
50
15
70
170
140
30
15
15
4X Ø16
100
15
70
15
70
60
100
60
50
60
30
30
30
30
70
60
60
10
36.2
30
37.6
30
36.2
10
EX-99
ON PCD Ø41
Ø2
A
Ø36
Ø46
Ø16
Ø46
Ø36
Ø16
Ø36
Ø36
5
5 + 5
20
5 + 5
40
SECTION A-A
A
Ø46
Ø36
Ø2
5 + 5
5 + 5
20
40
5 + 5
Ø36
Ø46
P-51

EX-100

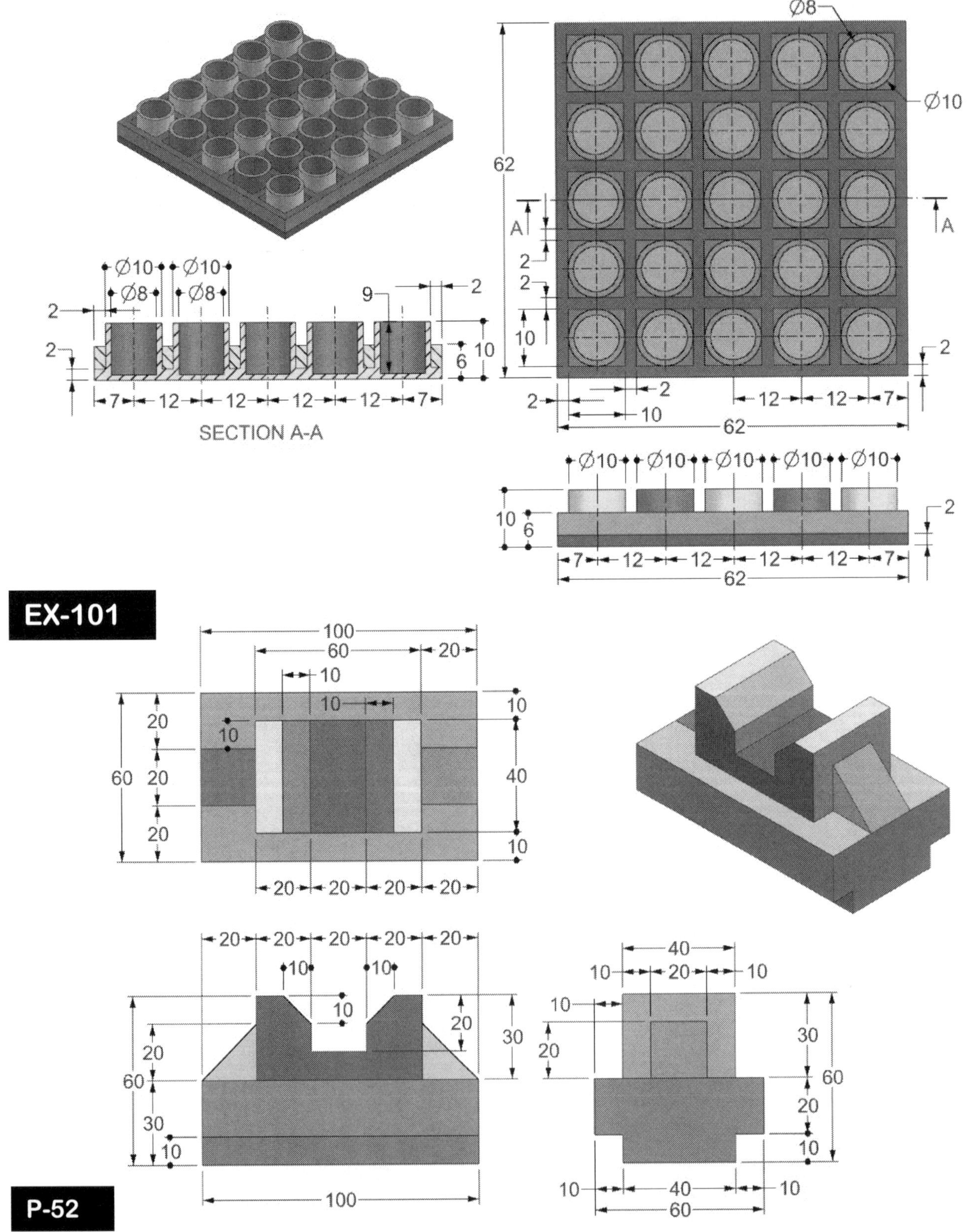

SECTION A-A

EX-101

P-52

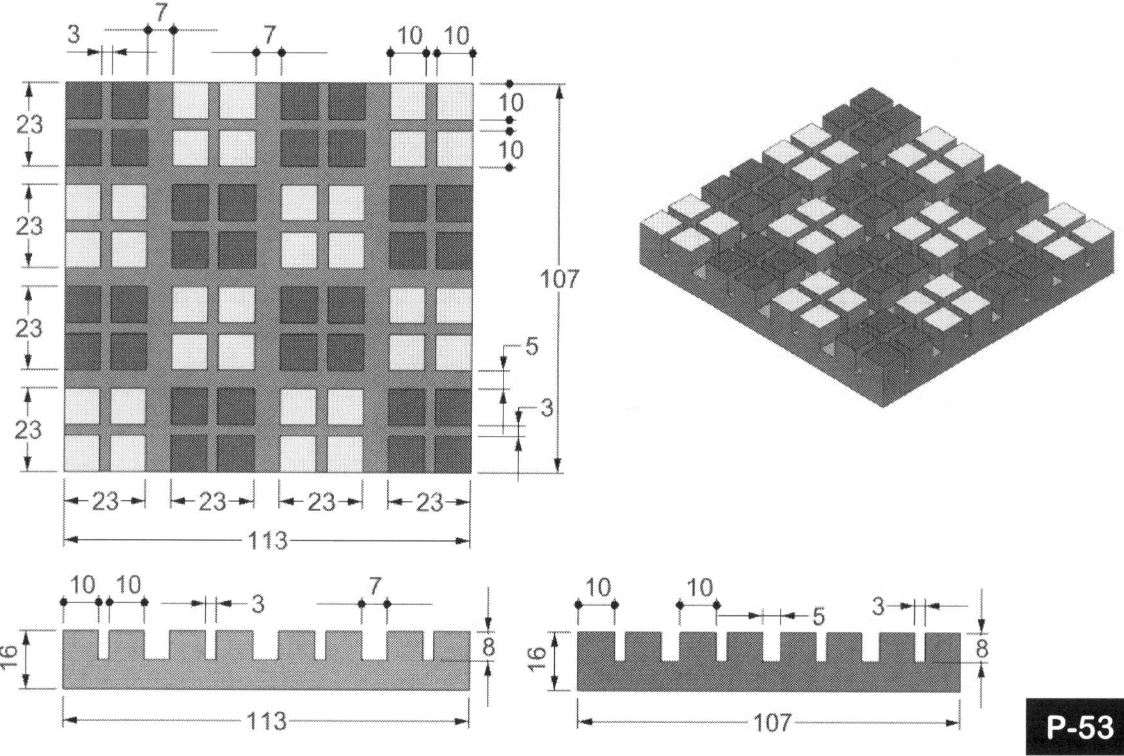

EX-102
124 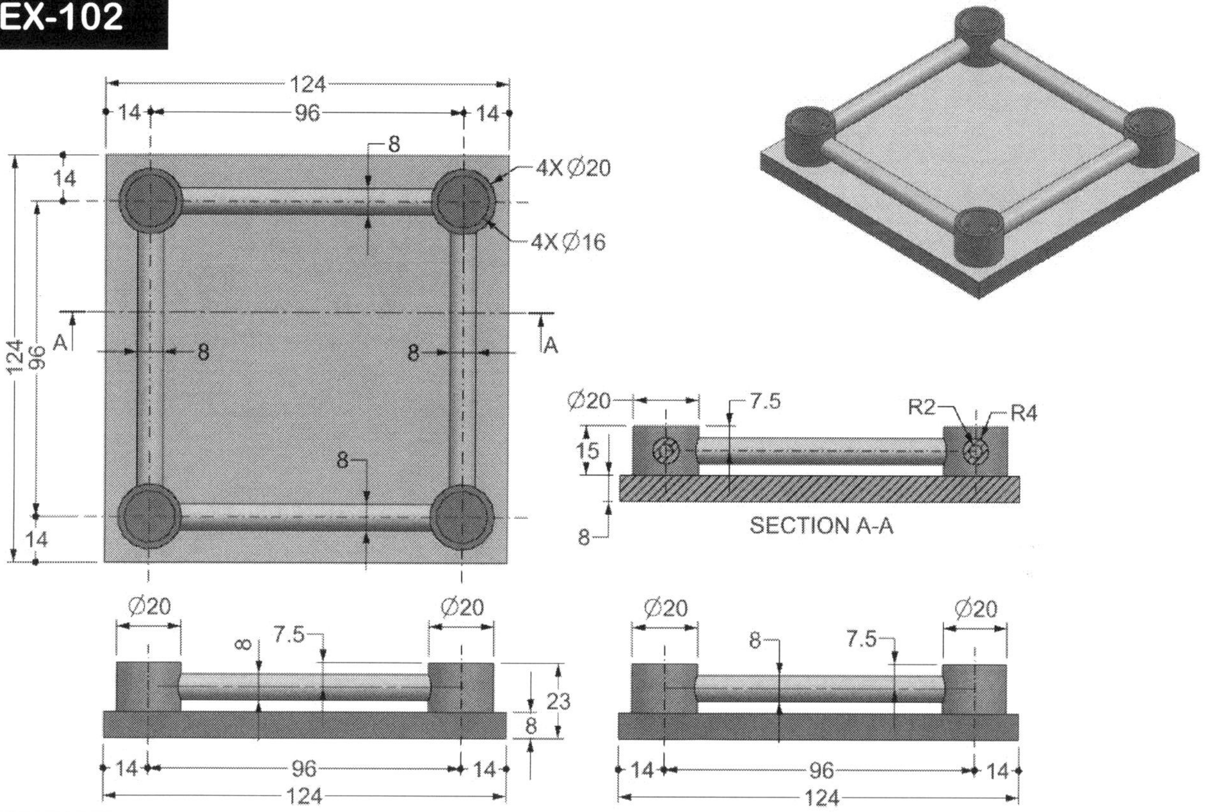
96
14
14
14
14
8
4X Ø20
4X Ø16
124
96
A
8
8
A
8
Ø20
7.5
R2
R4
15
8
SECTION A-A
Ø20
Ø20
8
7.5
23
8
14
96
14
124
Ø20
Ø20
8
7.5
14
96
14
124
EX-103
3
7
7
10
10
23
10
10
23
23
5
23
3
23
23
23
23
23
113
10
10
3
7
16
8
113
10
10
5
3
16
8
107
P-53

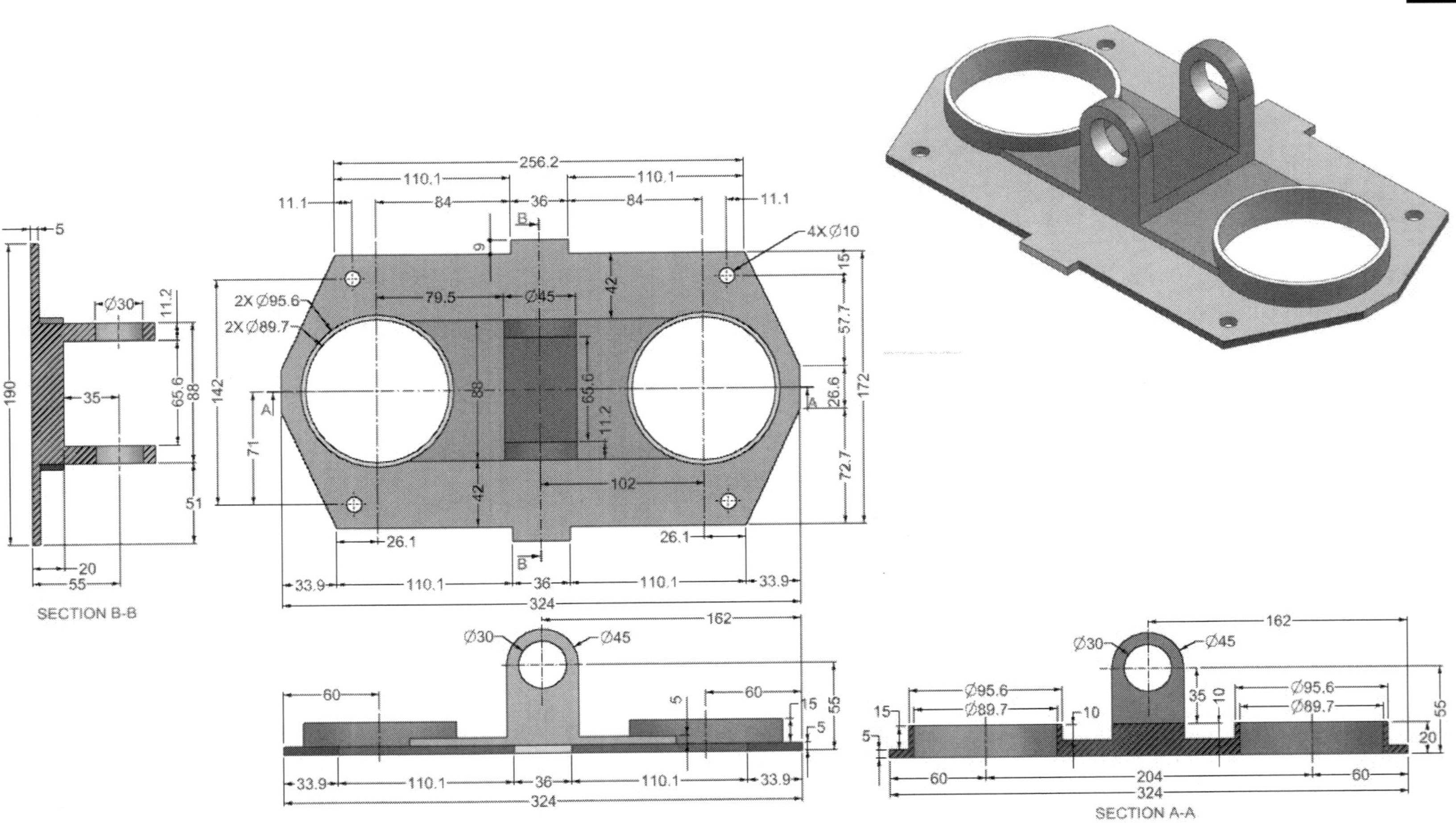

SECTION B-B
SECTION A-A
4X Ø10
2X Ø95.6
2X Ø89.7
Ø30
Ø45
Ø95.6
Ø89.7

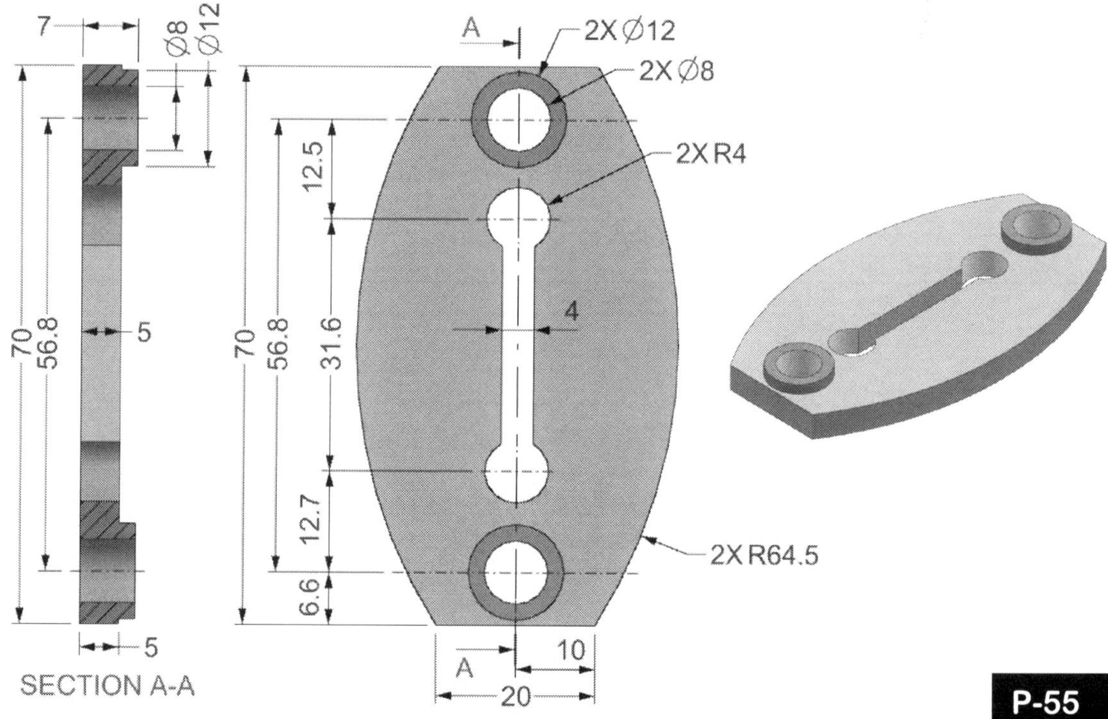

EX-105 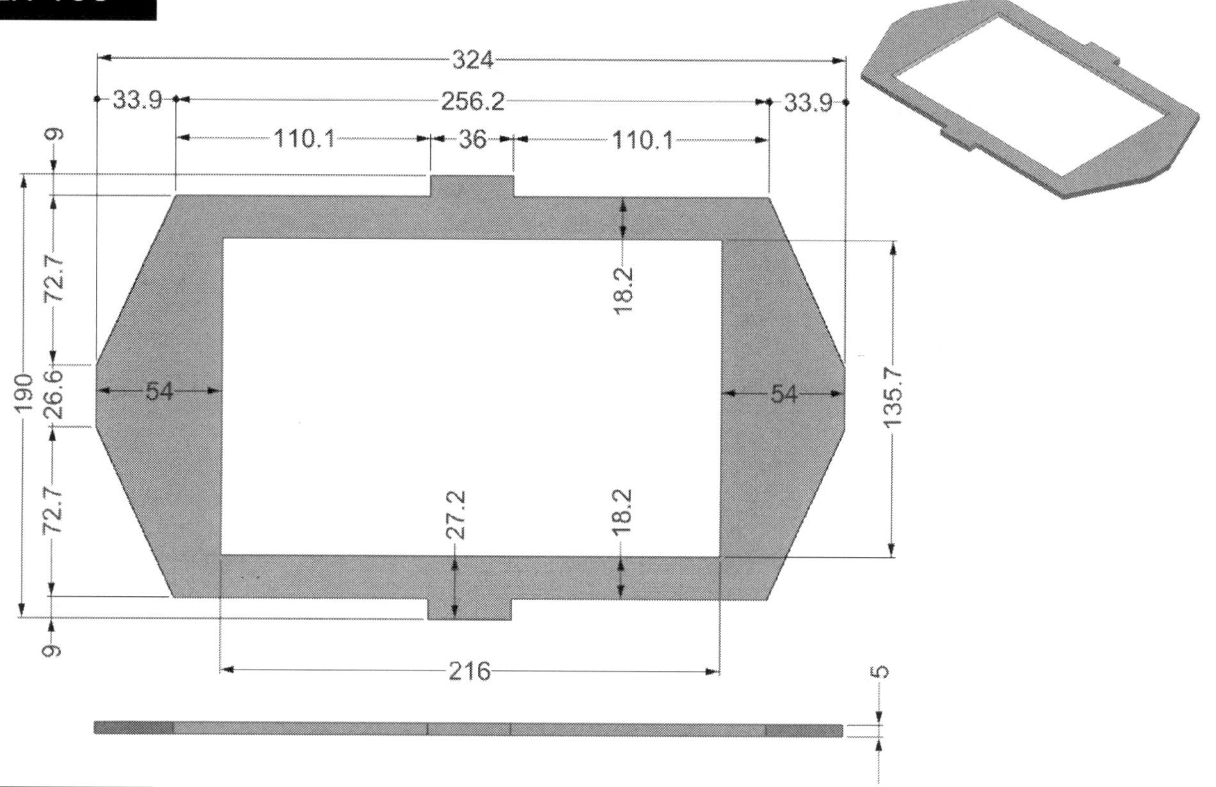
324
33.9
256.2
33.9
9
110.1
36
110.1
72.7
18.2
190
26.6
54
54
135.7
72.7
27.2
18.2
9
216
5
EX-106
7
Ø8
Ø12
A
2X Ø12
2X Ø8
12.5
2X R4
70
56.8
5
70
56.8
31.6
4
12.7
2X R64.5
6.6
A
10
SECTION A-A
A
20
P-55

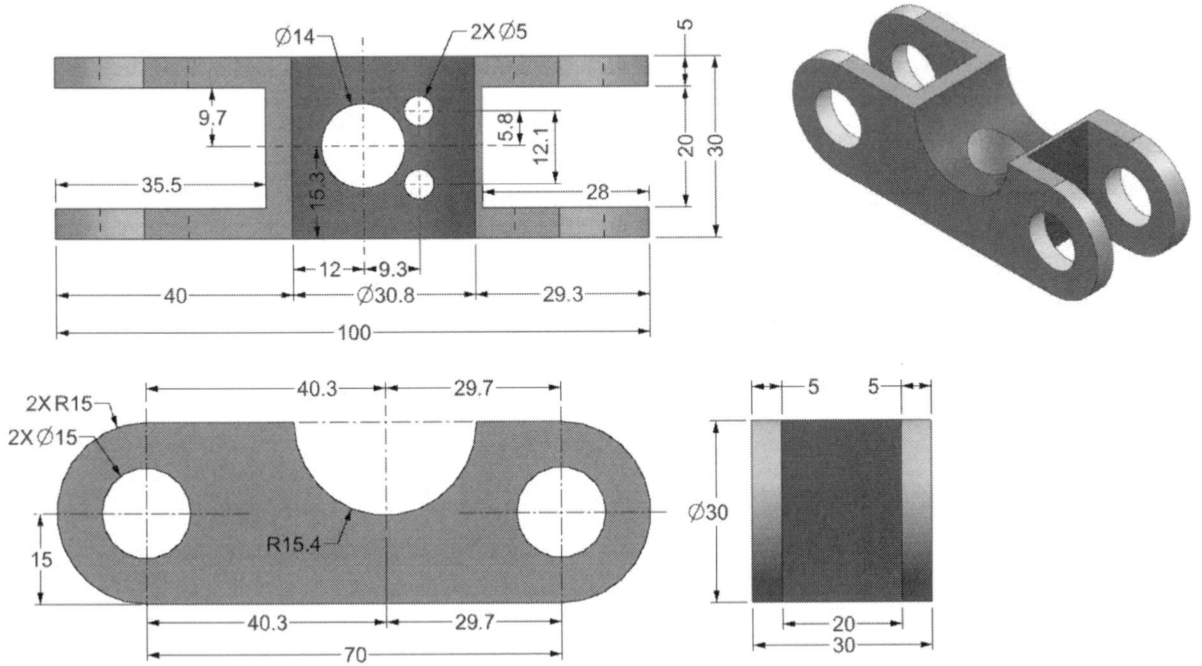

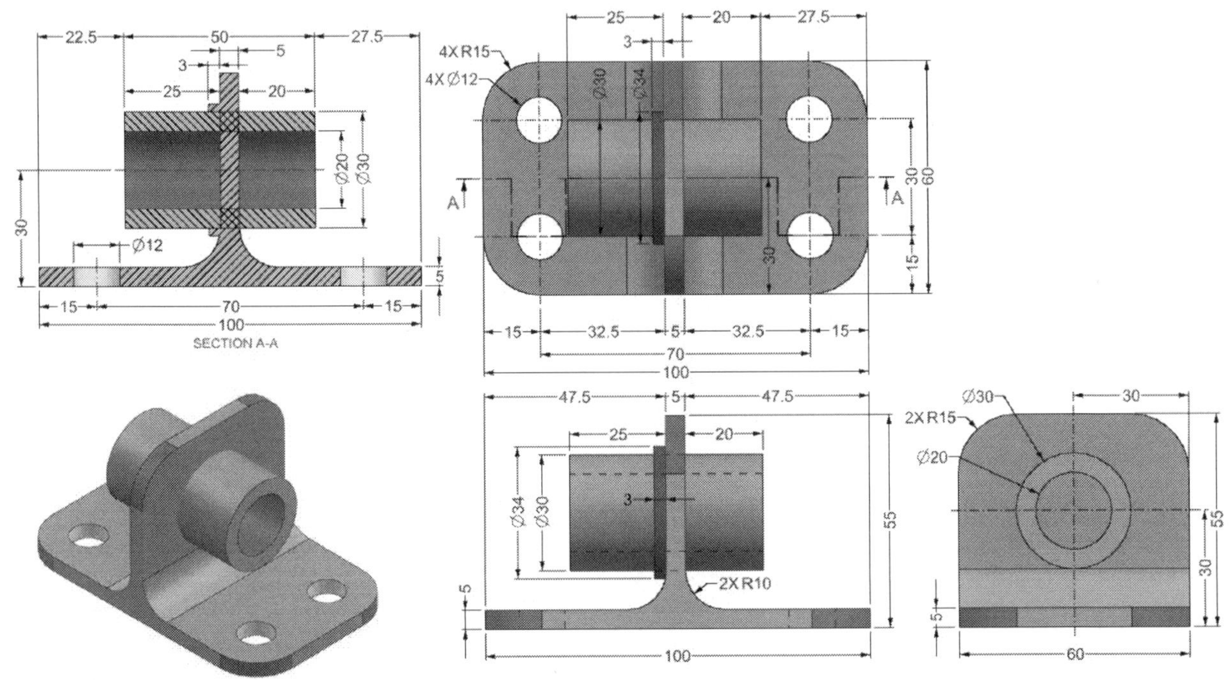

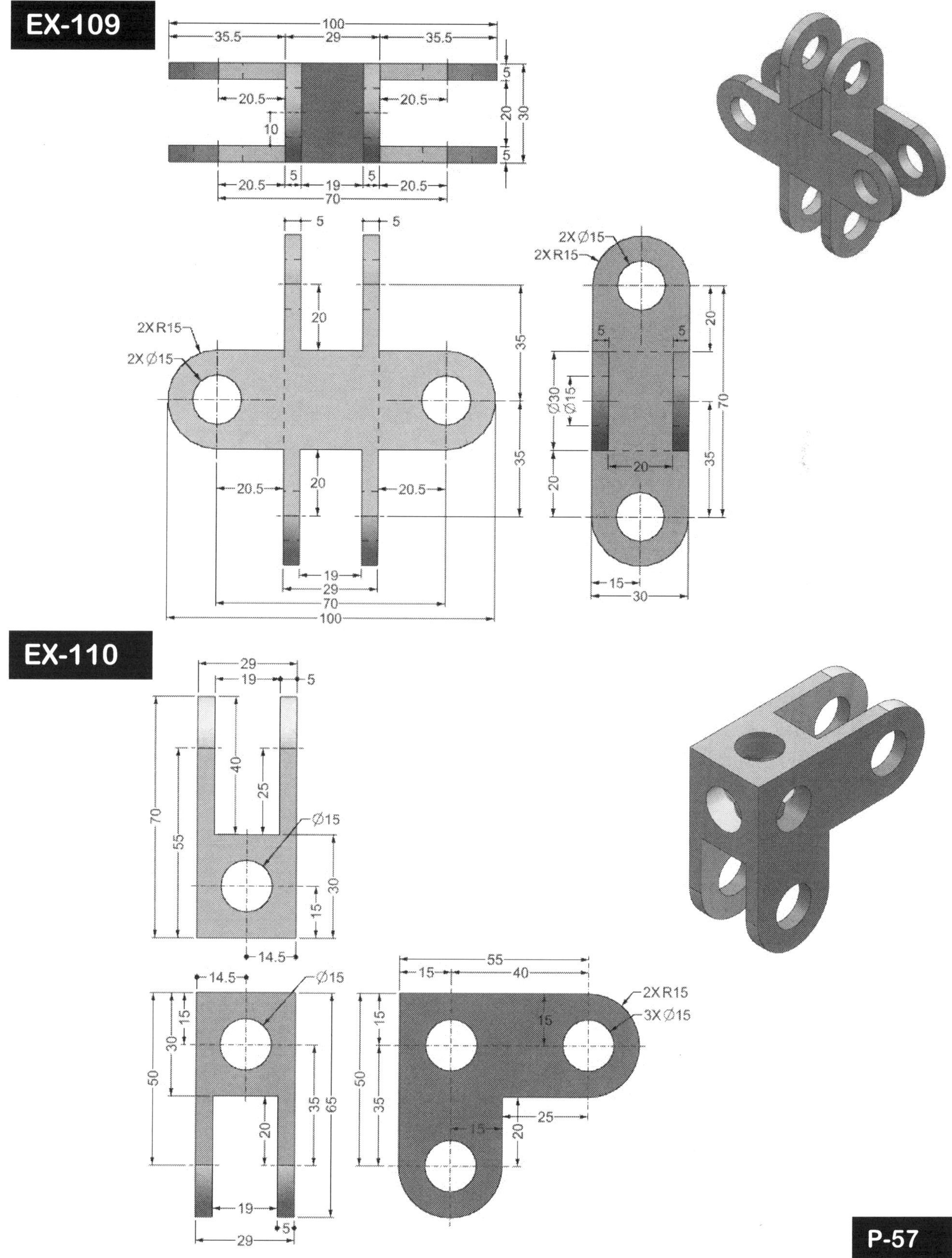

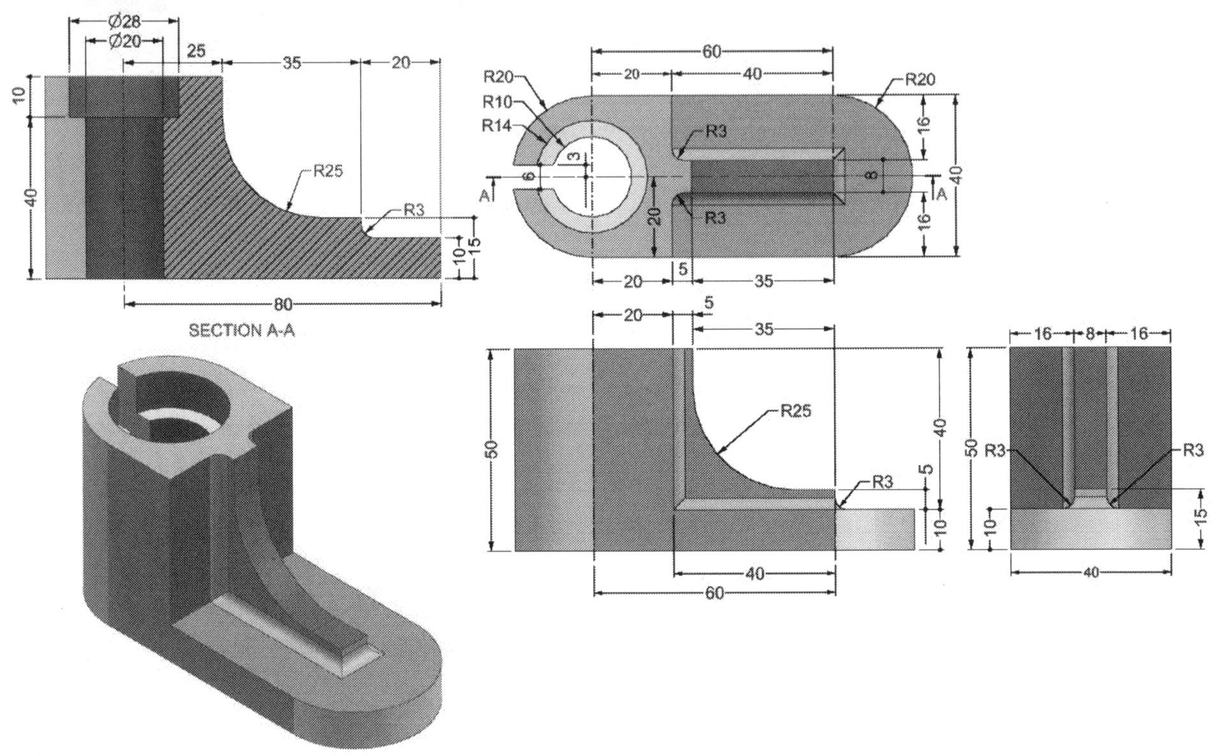

Ø28
Ø20
25
35
20
10
40
R25
R3
10
15
80
SECTION A-A
R20
R10
R14
60
20
40
R20
R3
16
3
6
A
20
R3
8
A
40
16
20
5
35
20
5
35
R25
50
40
R3
5
60
40
10
16
8
16
50
R3
R3
10
15
40

2X R10
2X R15
4X Ø14
4X Ø12
15
10
10
10
10
10
20
15
10
17.5
32.5
32.5
17.5
65
100
35
Ø30
35
Ø14
40
10
12
17.5
32.5
32.5
17.5
65
100
20
10
20
30
Ø14
Ø14
10
12
10
30
10
50

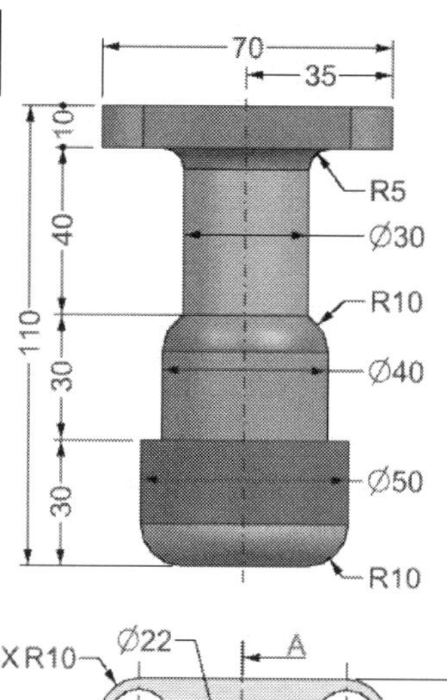

70
35
10
40
110
30
30
R5
Ø30
R10
Ø40
Ø50
R10

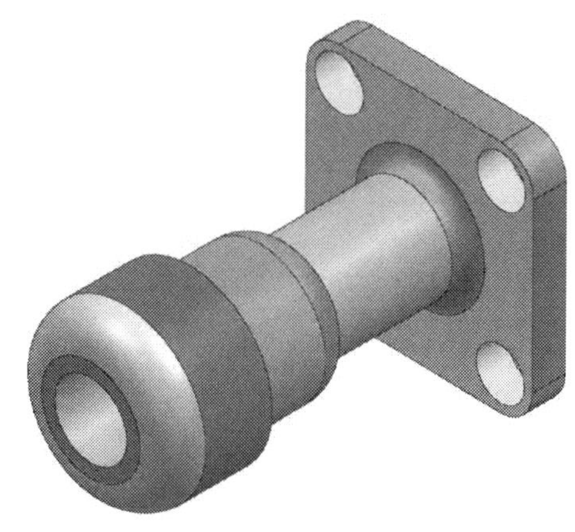

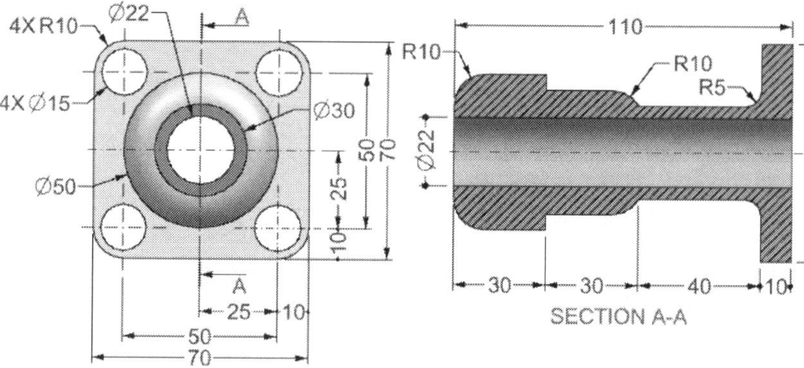

4X R10
Ø22
A
Ø30
4X Ø15
Ø50
A
25
10
50
70
50
70
25
10
R10
R10
R5
110
Ø22
70
30
30
40
10
SECTION A-A

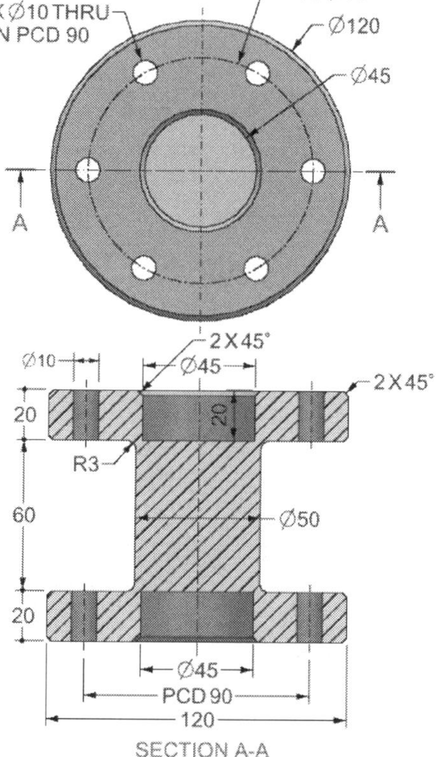

6X Ø10 THRU
ON PCD 90
PCD Ø90
Ø120
Ø45
A
A
2 X 45°
Ø10
Ø45
20
20
2 X 45°
R3
60
Ø50
20
Ø45
PCD 90
120
SECTION A-A

P-60

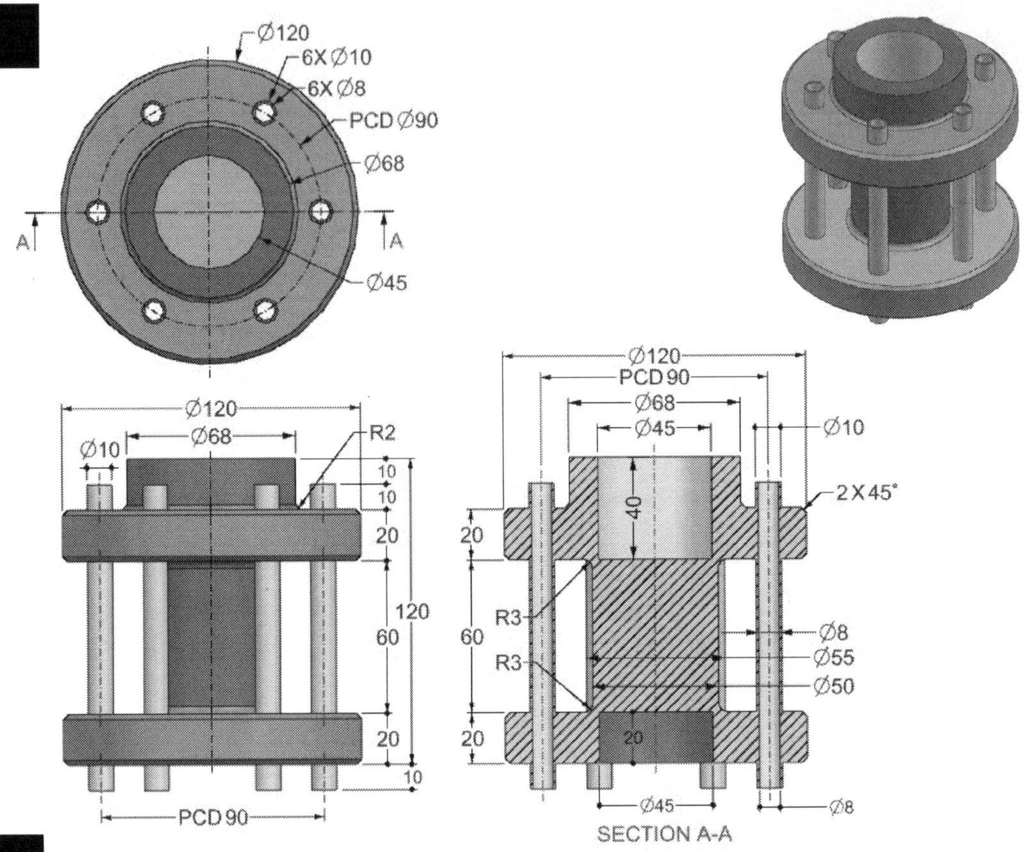

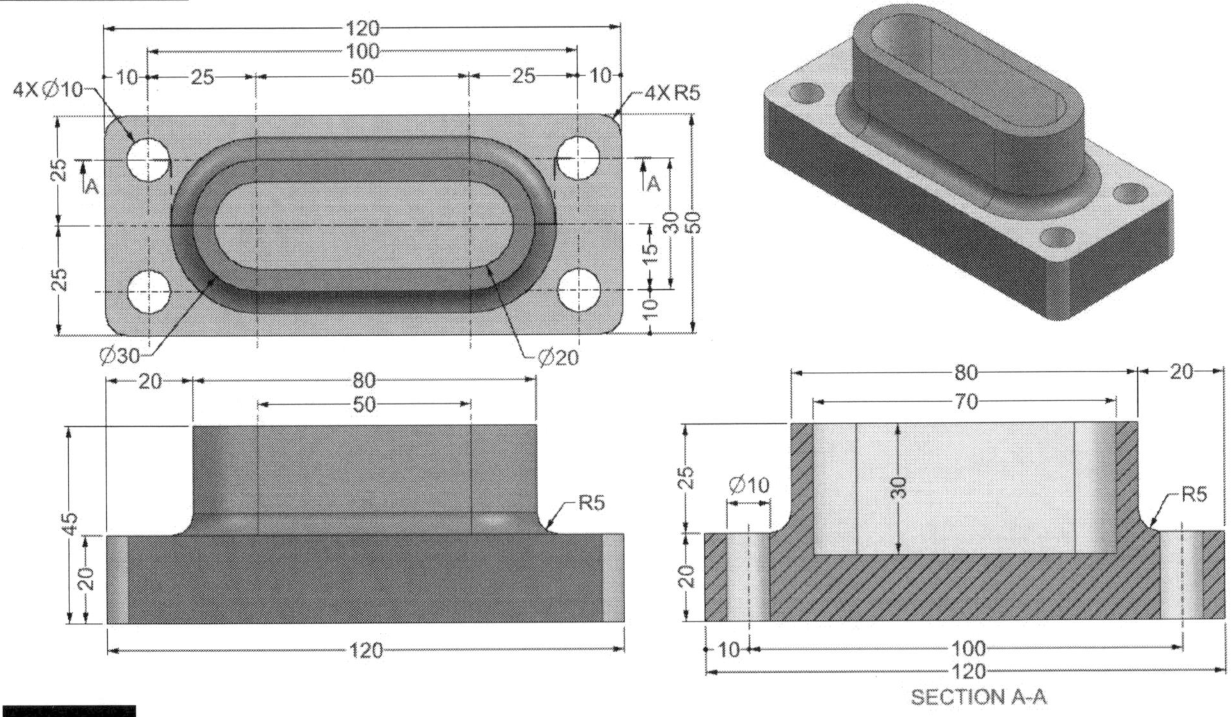

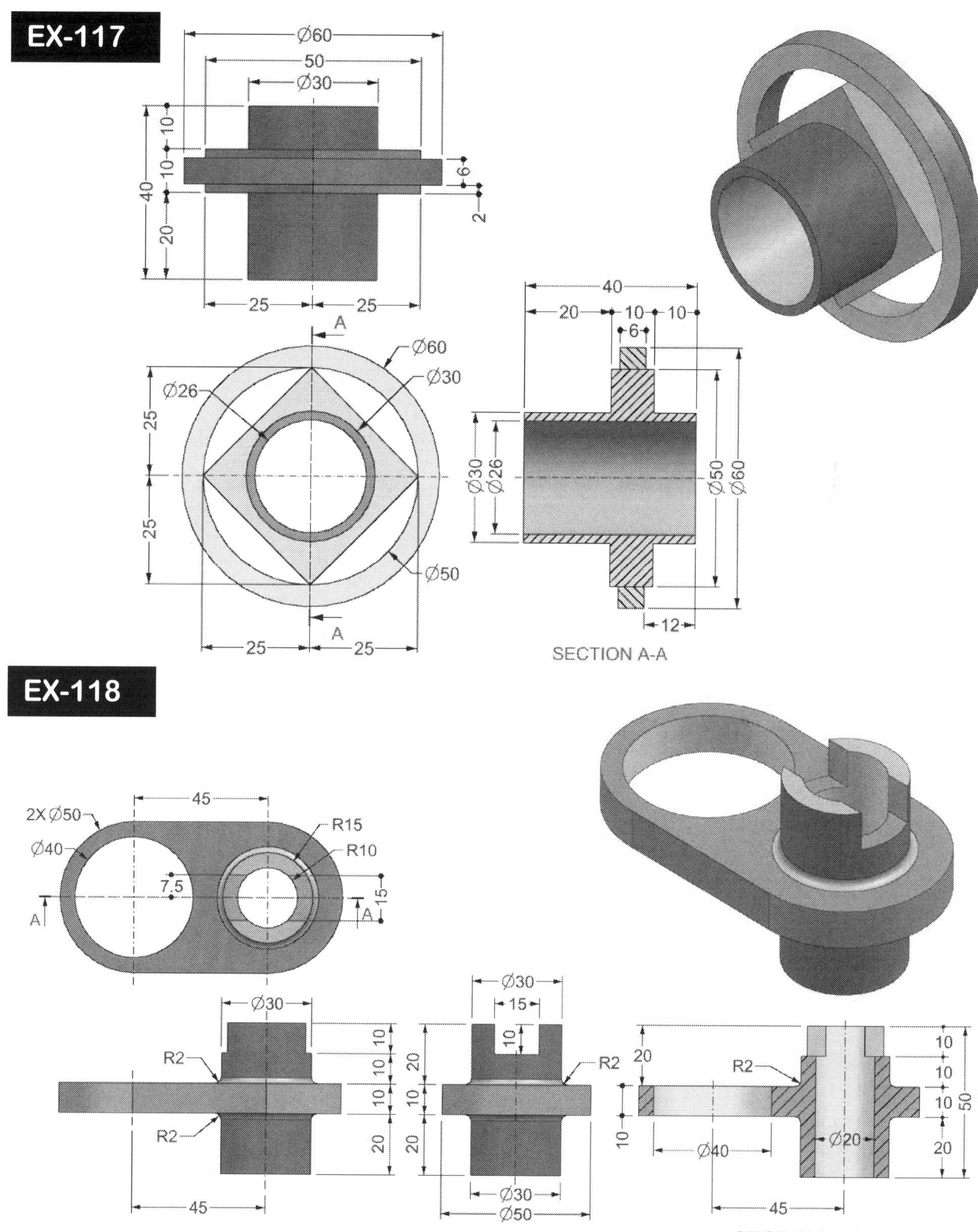

EX-117
Ø60
50
Ø30
10
10
40
20
6
2
25
25
A
Ø60
Ø26
Ø30
25
25
Ø50
25
25
A
40
20
10
10
6
Ø30
Ø26
Ø50
Ø60
12
SECTION A-A

EX-118
2X Ø50
Ø40
R15
R10
45
7.5
15
A
A
Ø30
R2
10
10
10
20
R2
10
10
20
20
45
Ø30
15
10
20
10
R2
Ø30
Ø50
Ø30
20
R2
10
20
10
Ø40
Ø20
10
10
10
20
50
45
SECTION A-A

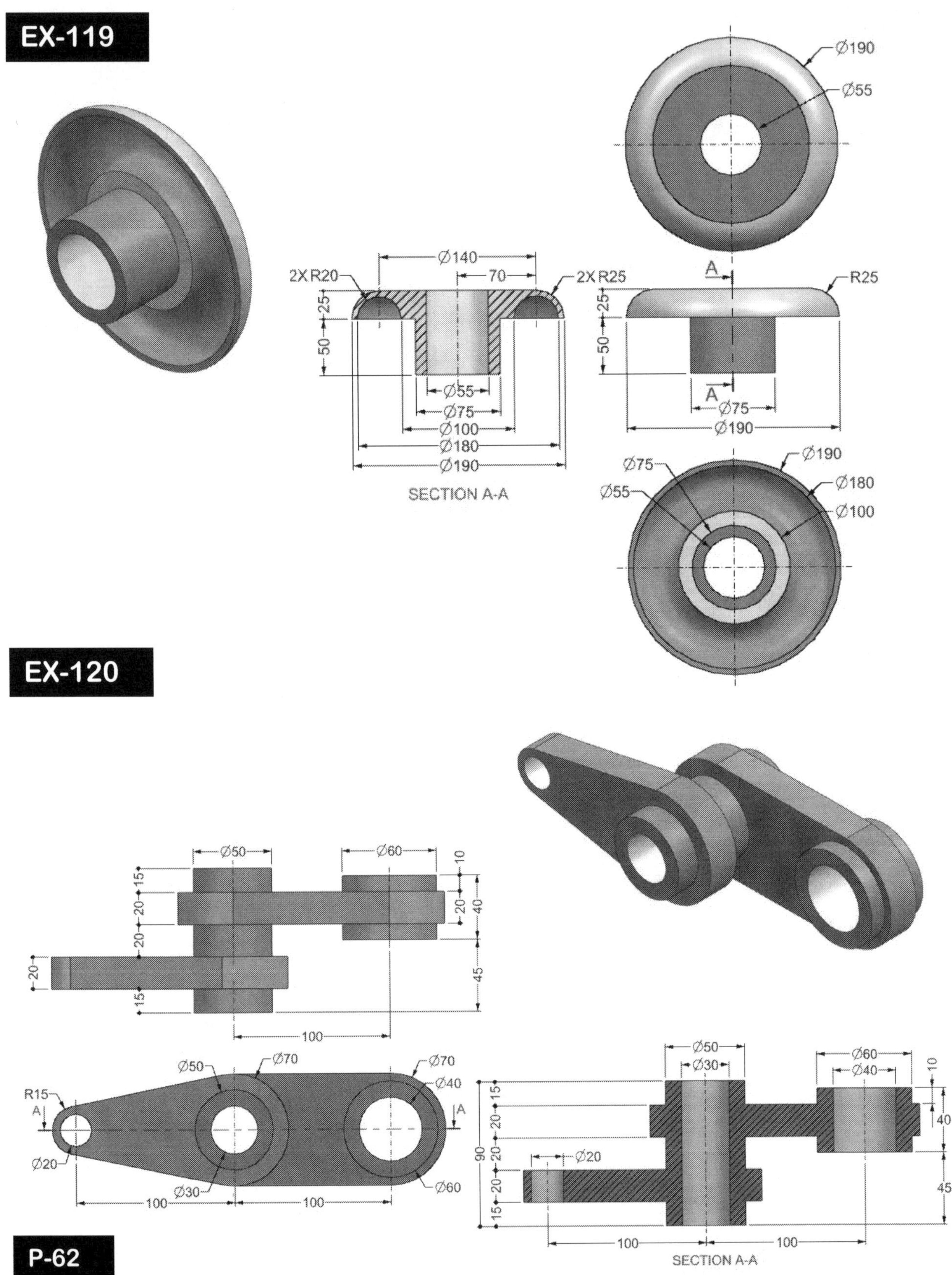

EX-119
Ø190
Ø55
Ø140
70
2X R20
2X R25
25
50
Ø55
Ø75
Ø100
Ø180
Ø190
SECTION A-A
A
25
50
R25
Ø75
Ø190
A
Ø75
Ø190
Ø55
Ø180
Ø100
EX-120
Ø50
Ø60
15
20
20
10
20
40
20
45
15
100
R15
A
Ø50
Ø70
Ø70
Ø40
A
Ø20
Ø30
Ø60
100
100
Ø50
Ø30
Ø60
Ø40
15
20
20
10
90
Ø20
40
20
15
100
100
45
SECTION A-A
P-62

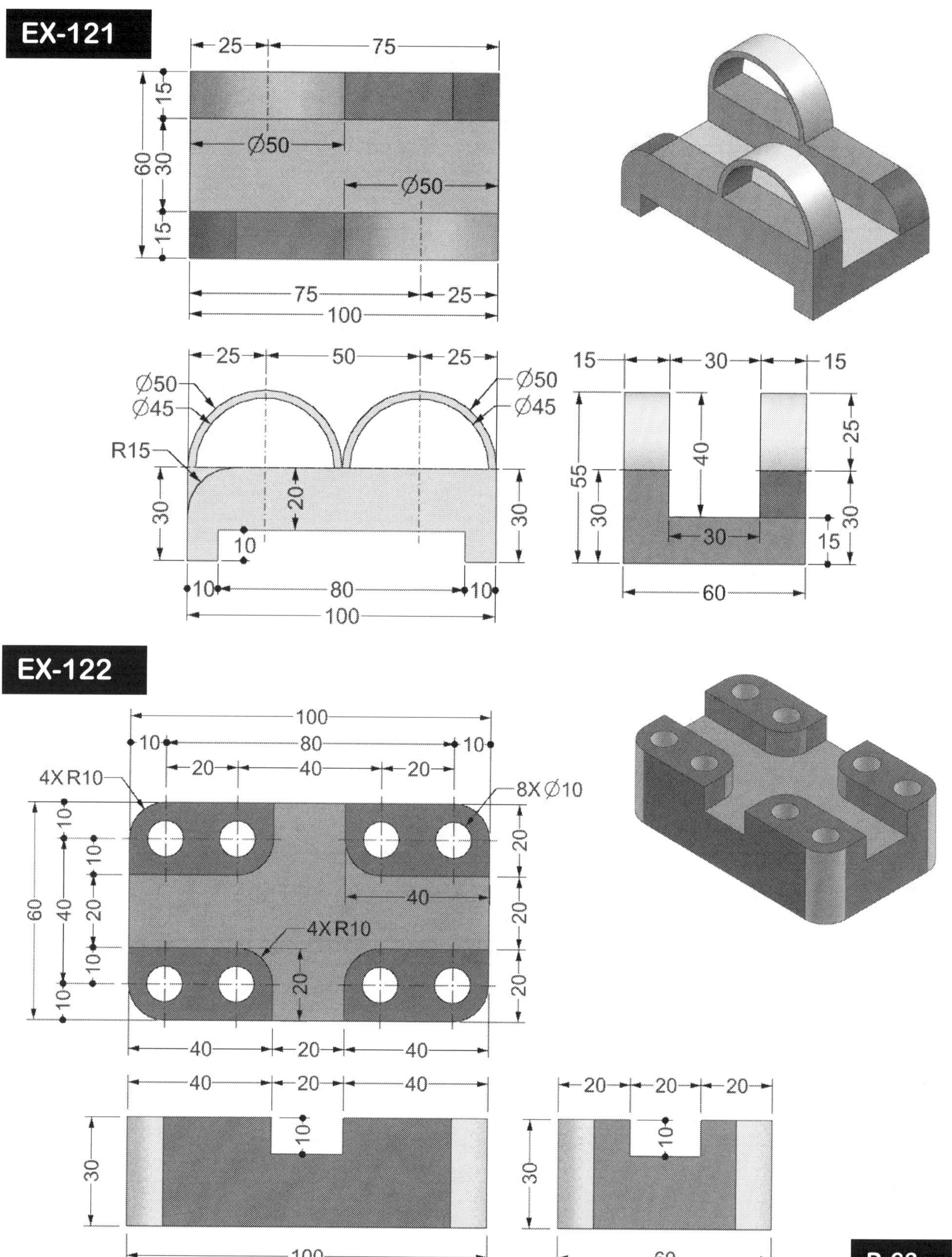

EX-121
25
75
15
60
30
15
Ø50
Ø50
75
25
100
25
50
25
Ø50
Ø45
Ø50
Ø45
R15
30
20
30
10
10
80
10
100
15
30
15
55
30
40
25
30
30
15
60
EX-122
100
10
80
10
20
40
20
4XR10
8X Ø10
10
40
20
10
10
20
10
60
20
20
40
20
20
4XR10
40
20
40
40
20
40
30
10
100
20
20
20
30
10
60
P-63

EX-123
10 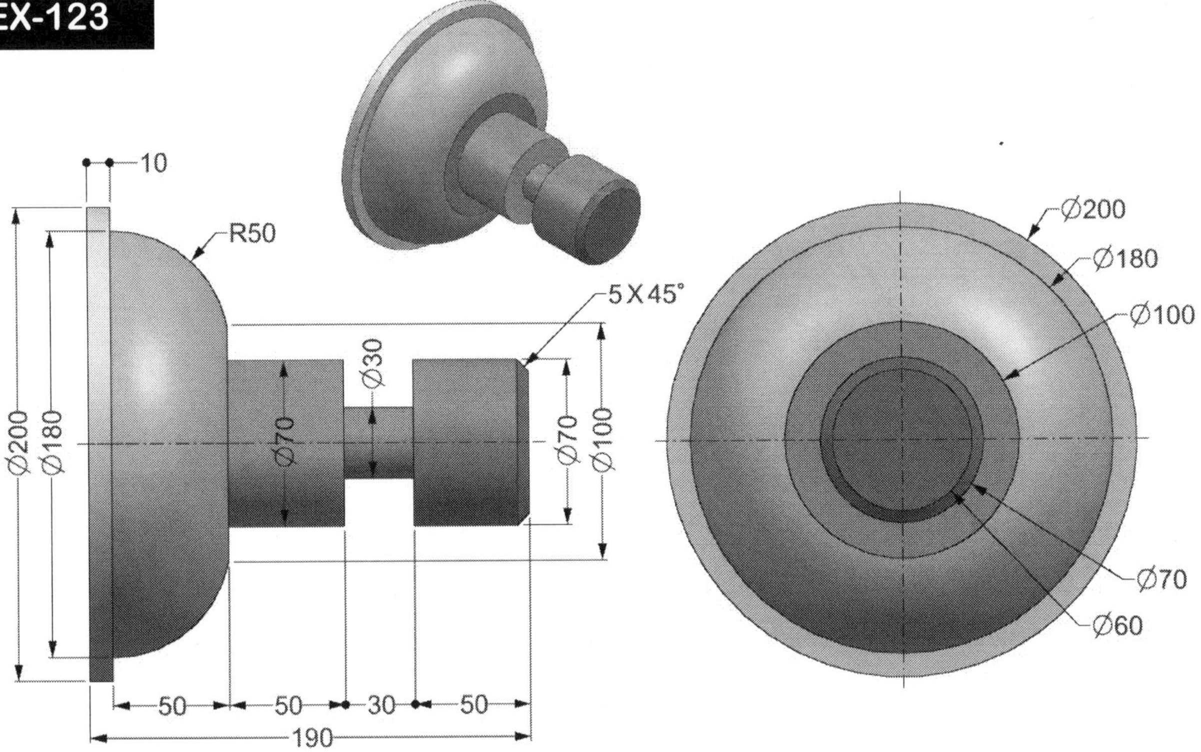
R50
5 X 45°
Ø30
Ø70
Ø70
Ø100
Ø200
Ø180
50
50
30
50
190
Ø200
Ø180
Ø100
Ø70
Ø60

EX-124
50 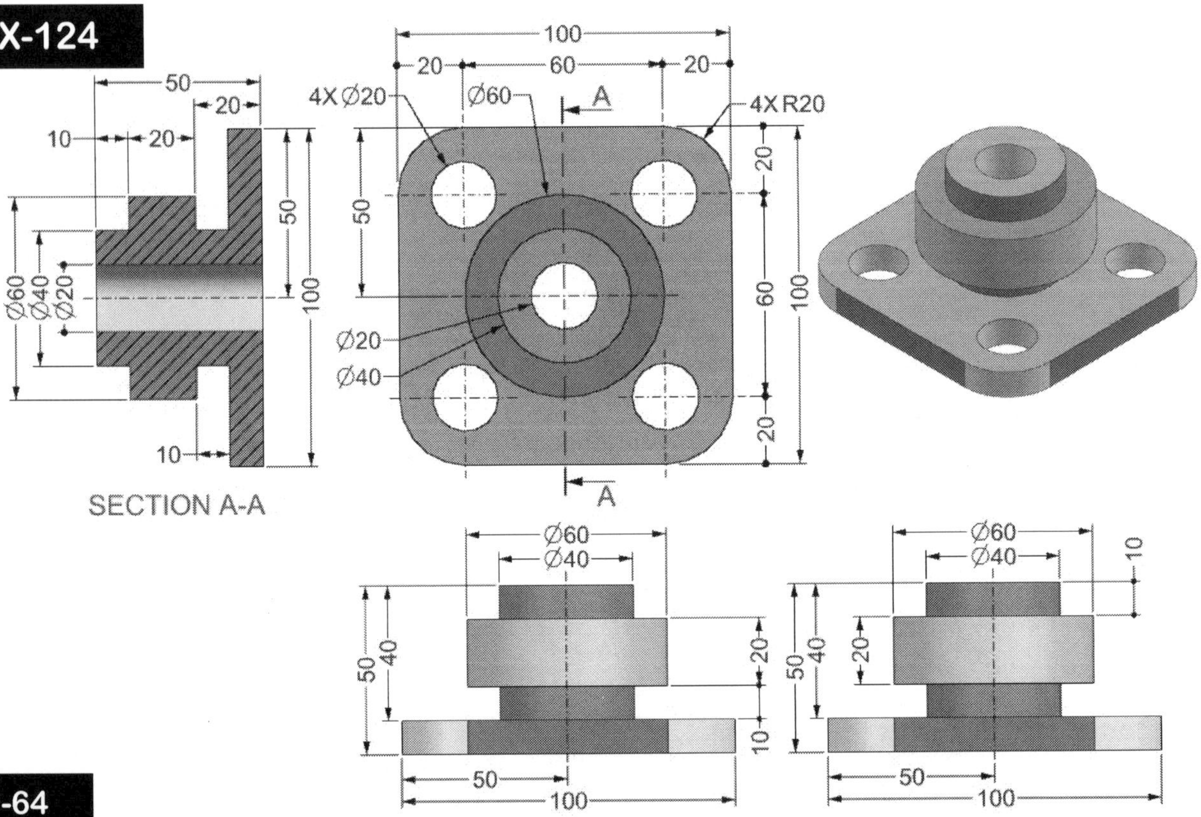
20
10
20
4X Ø20
Ø60
A
4X R20
100
20
60
20
20
50
60
100
Ø60
Ø40
Ø20
Ø20
Ø40
A
SECTION A-A
10
10
Ø60
Ø40
50
40
20
10
50
100
Ø60
Ø40
50
40
20
10
50
100

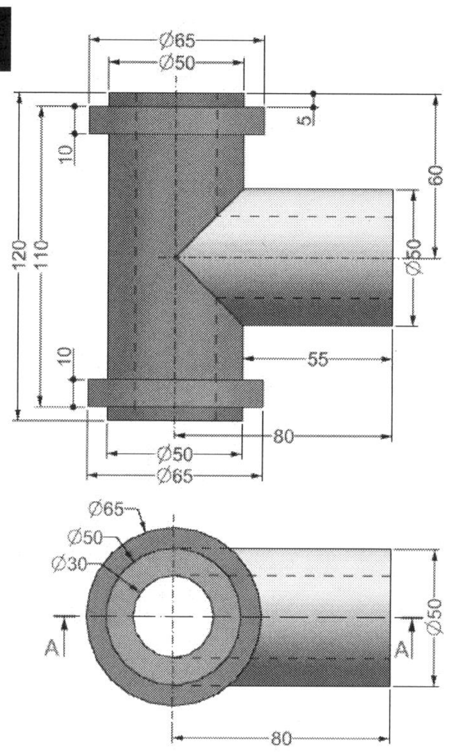

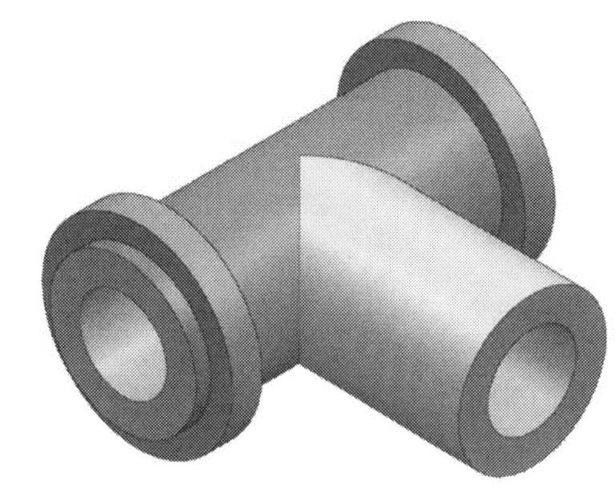

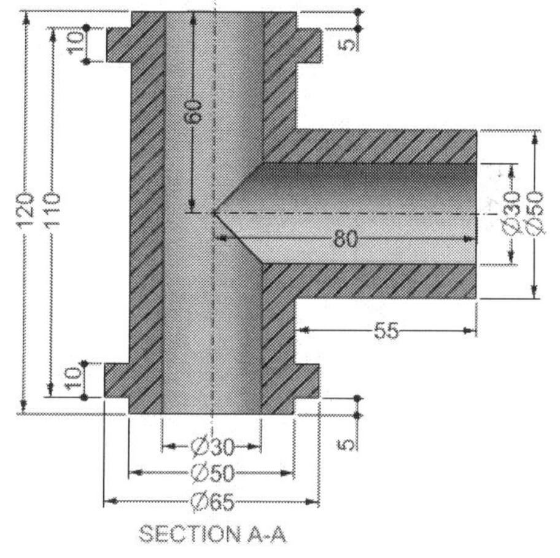

SECTION A-A

EX-126

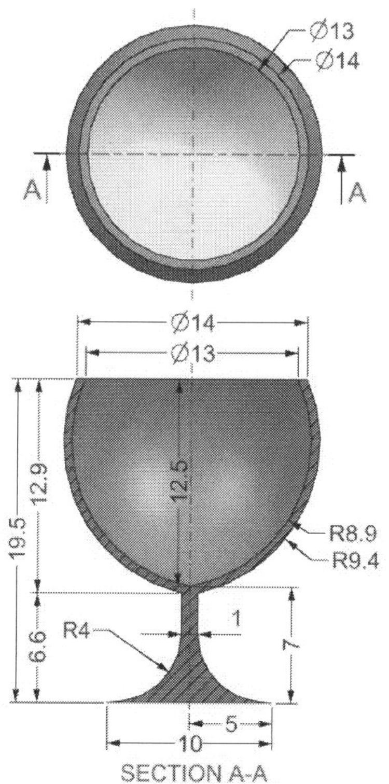

SECTION A-A

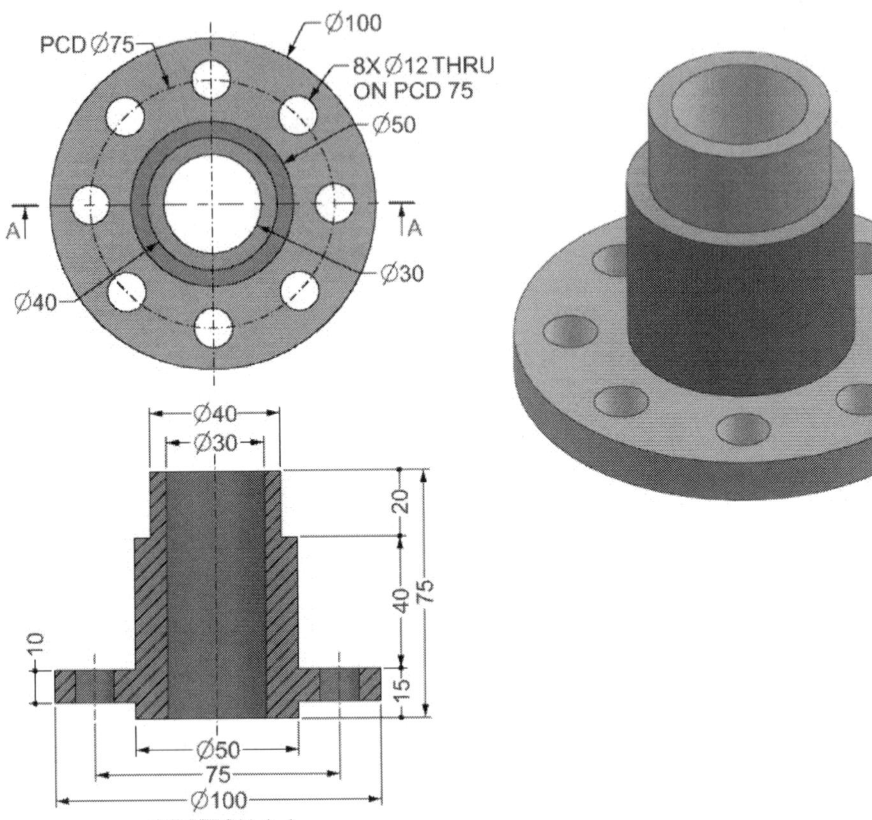

PCD Ø75
Ø100
8X Ø12 THRU
ON PCD 75
Ø50
Ø40
Ø30
A
A
Ø40
Ø30
20
40
75
10
15
Ø50
75
Ø100
SECTION A-A

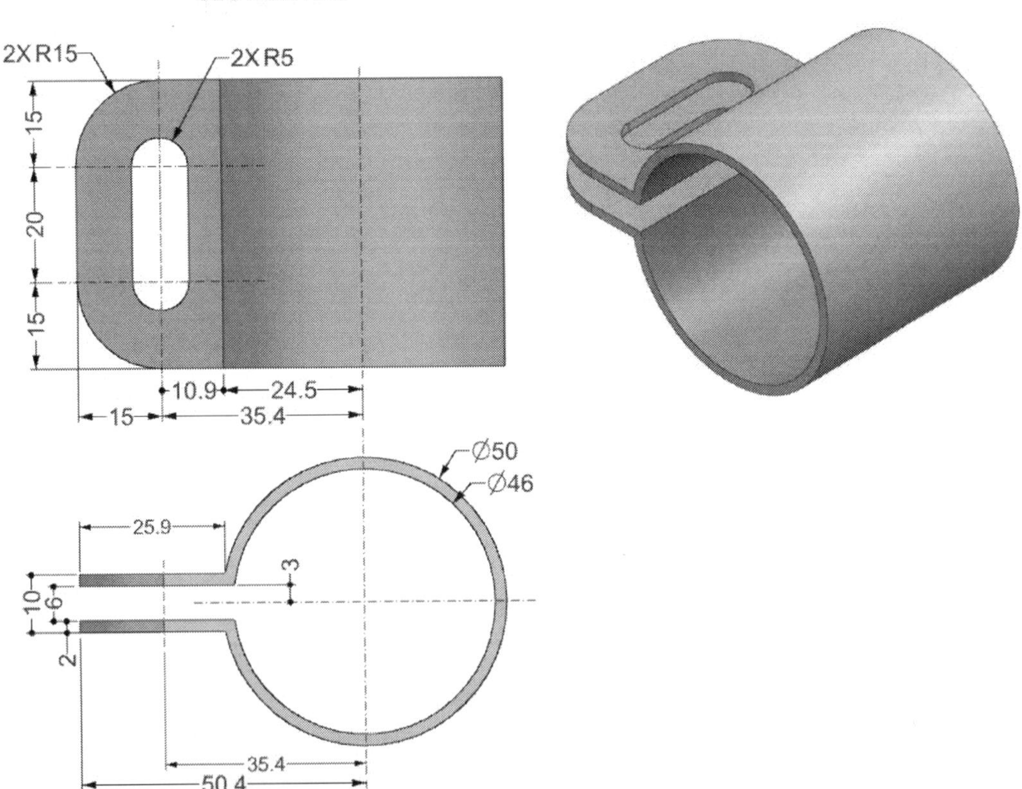

2X R15
2X R5
15
20
15
10.9
24.5
15
35.4
Ø50
Ø46
25.9
3
10
6
2
35.4
50.4

EX-129

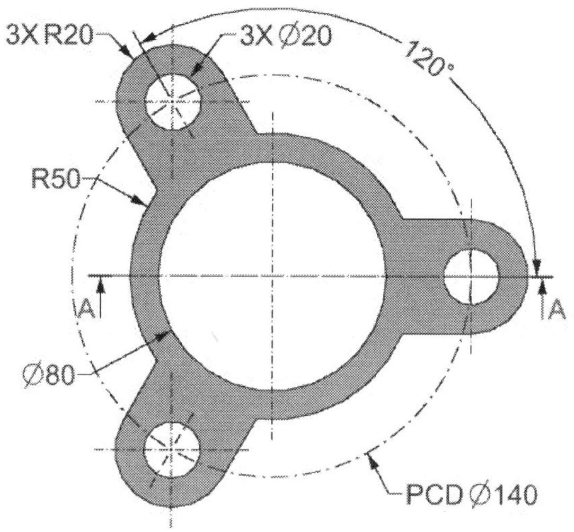

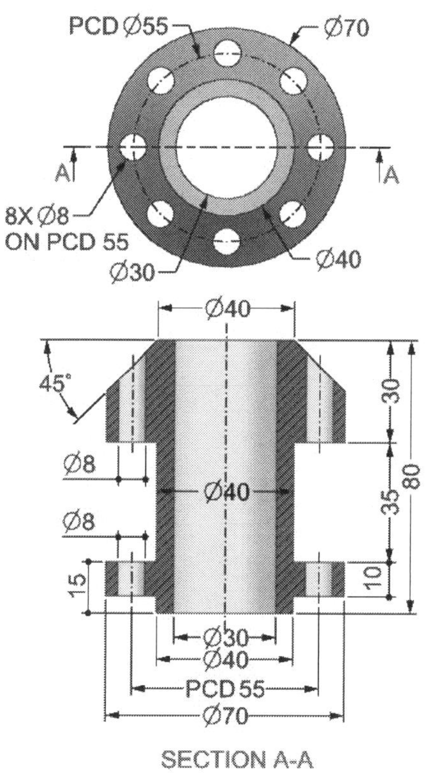

3X R20
3X Ø20
120°
R50
A
A
Ø80
PCD Ø140

Ø80
Ø20
15
Ø100
50
80
70
Ø20

SECTION A-A

EX-130

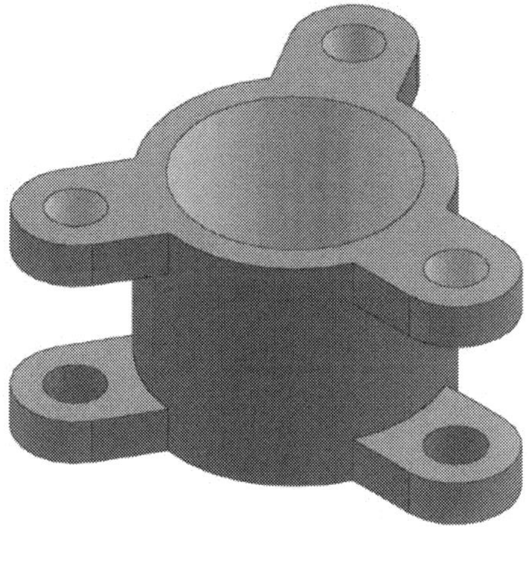

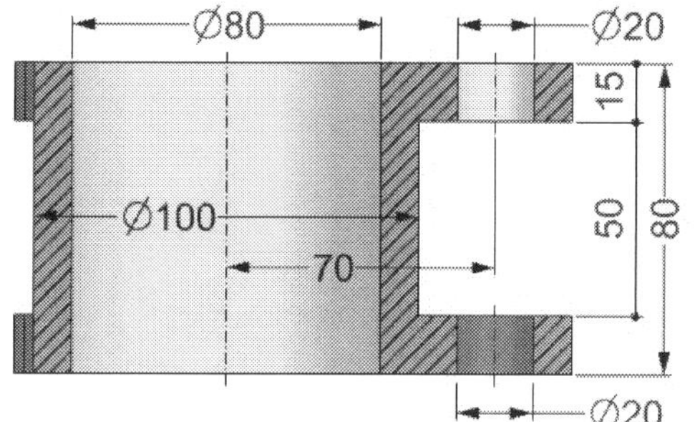

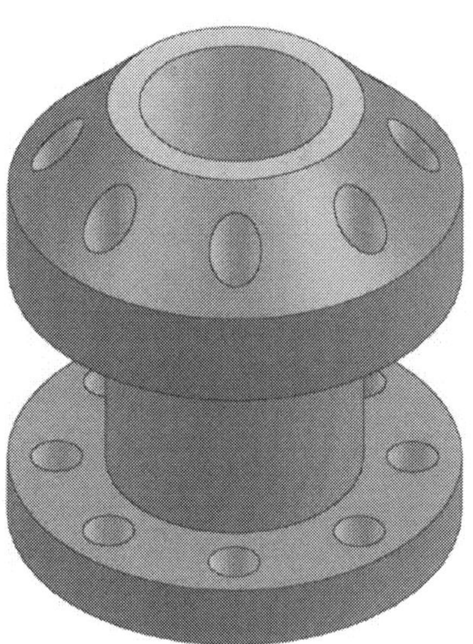

PCD Ø55
Ø70
A
A
8X Ø8
ON PCD 55
Ø30
Ø40

Ø40
45°
30
Ø8
Ø40
35
80
Ø8
15
10
Ø30
Ø40
PCD 55
Ø70

SECTION A-A

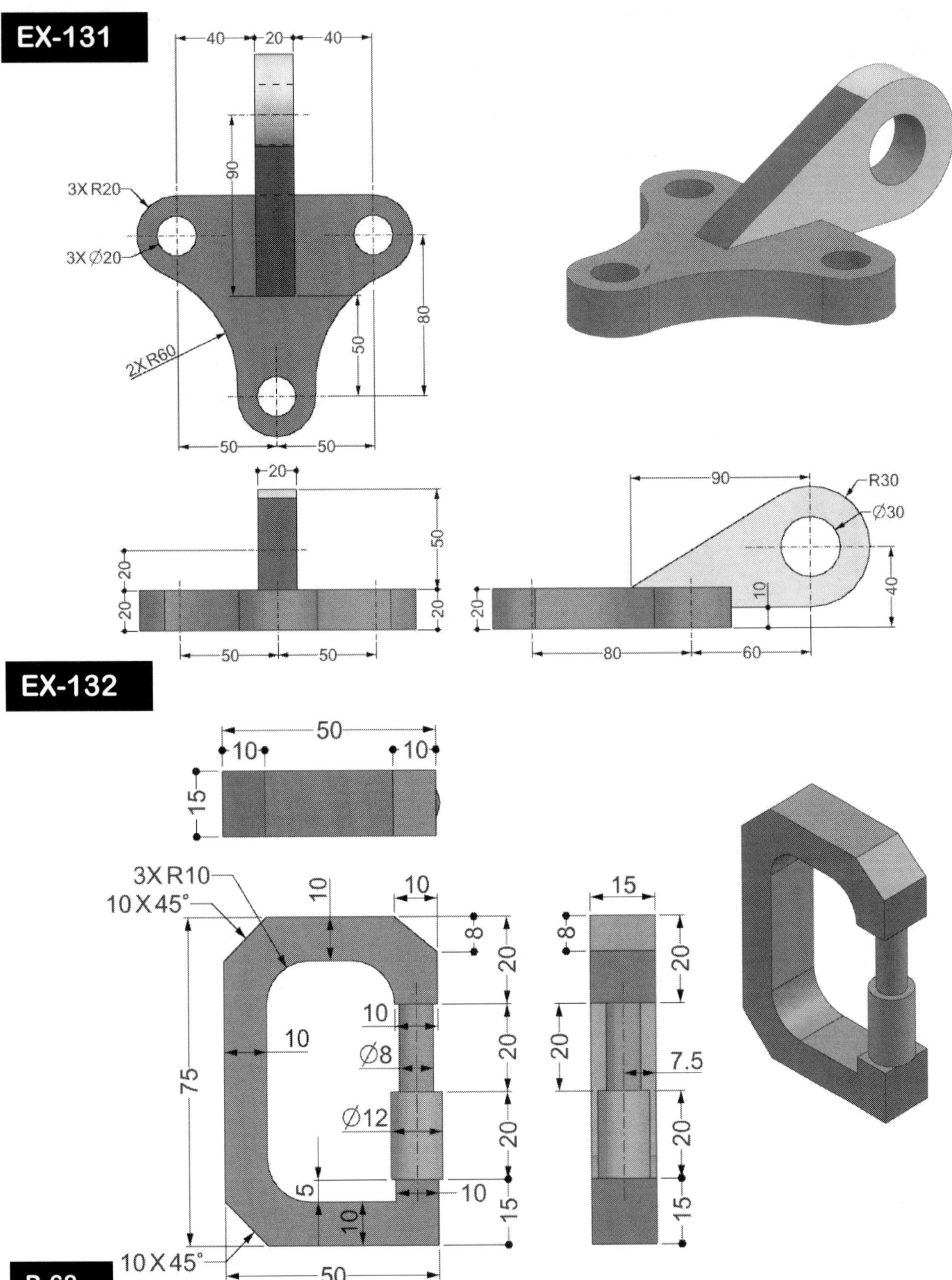

3X R20
3X Ø20
2X R60
40
20
40
90
80
50
50
50
20
20
20
50
50
20
50
90
R30
Ø30
10
40
20
80
60

50
10
10
15
3X R10
10 X 45°
10
10
10
8
20
10
Ø8
20
Ø12
20
5
10
10
75
50
15
15
20
8
20
20
7.5
20
15

EX-134

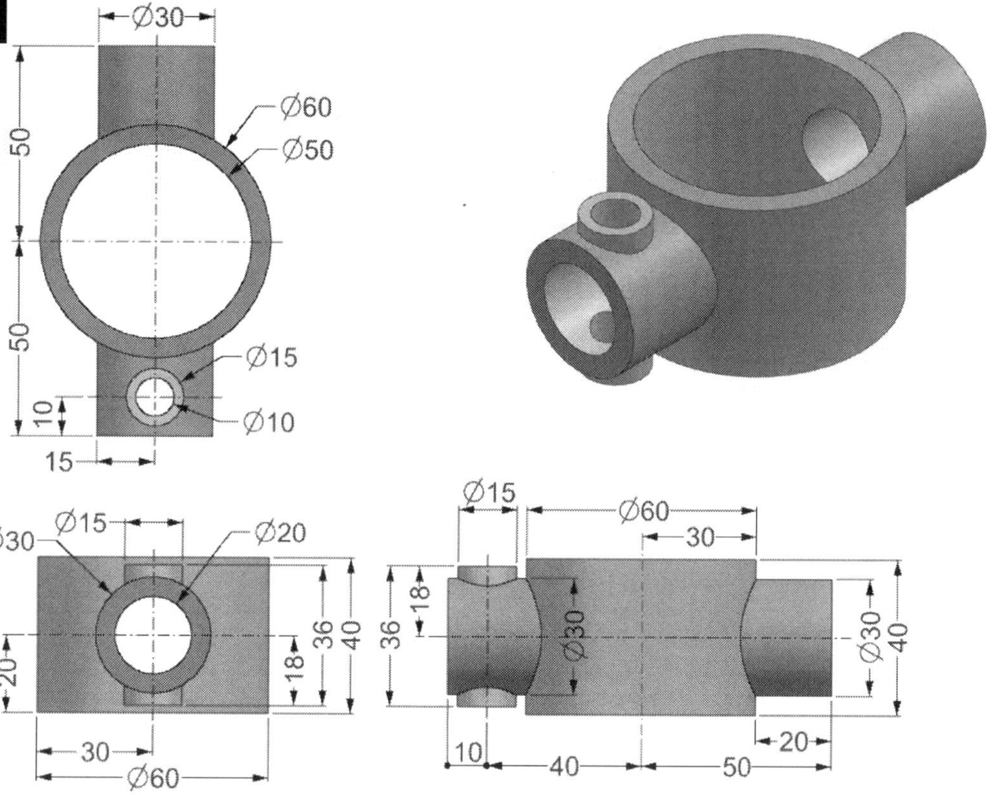

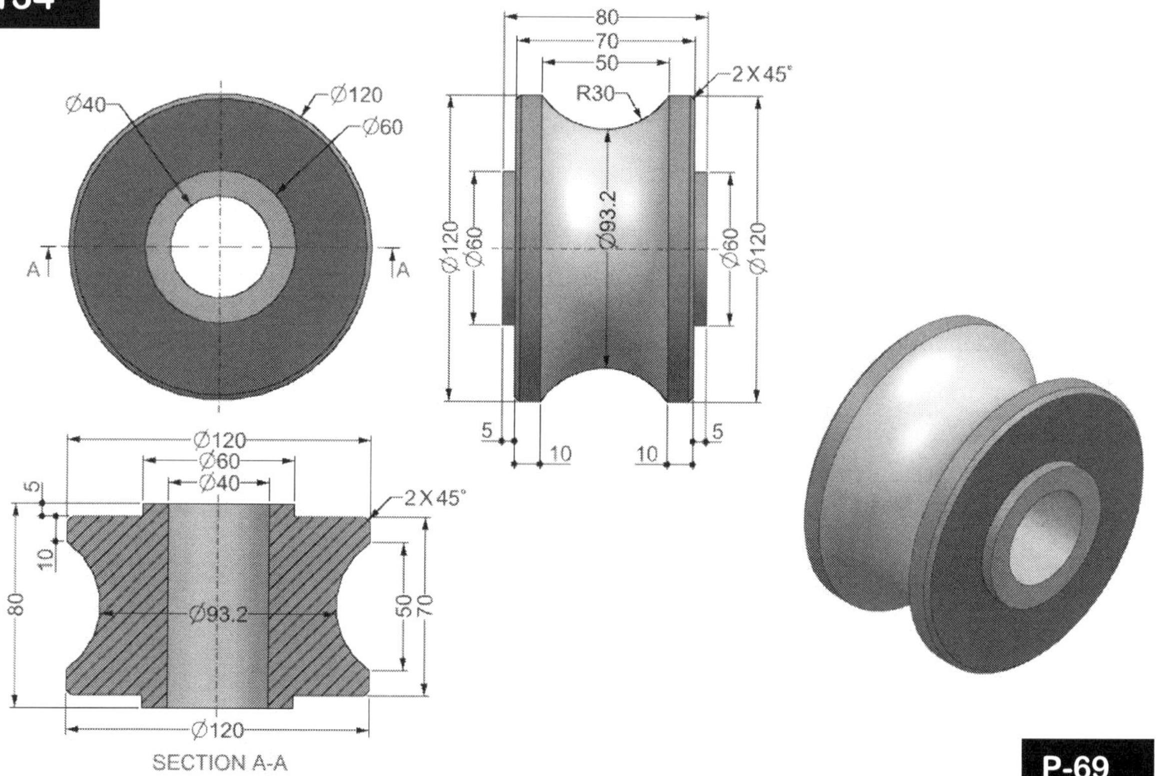

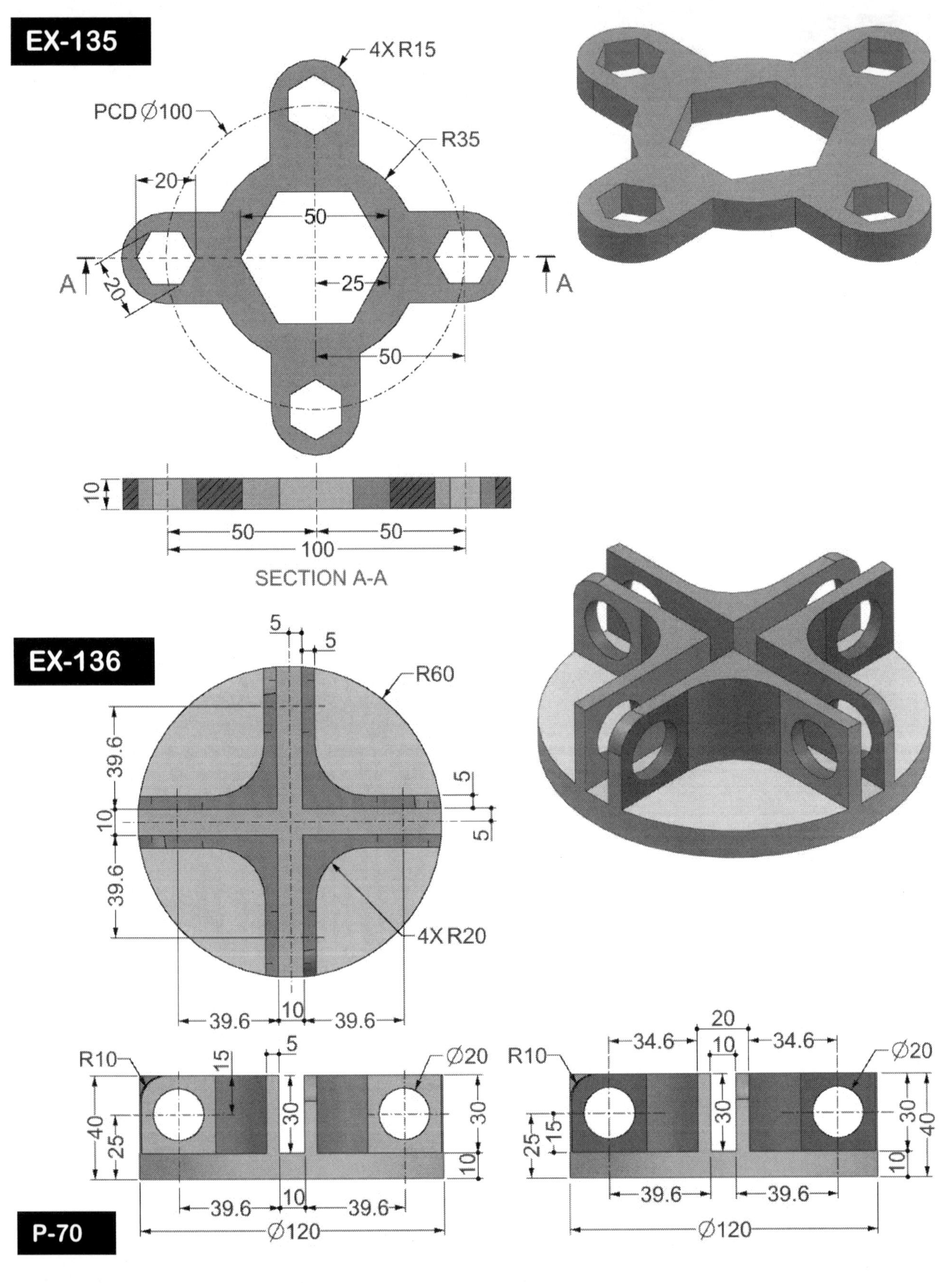
EX-135
4X R15
PCD Ø100
R35
20
50
25
50
20
A
A
10
50
50
100
SECTION A-A
EX-136
5
5
R60
39.6
39.6
10
5
5
39.6
10
39.6
4X R20
R10
15
5
Ø20
40
25
30
30
10
39.6
10
39.6
Ø120
R10
20
34.6
10
34.6
Ø20
25
15
30
30
40
10
39.6
39.6
Ø120
P-70

2X R6
2X R5
20
A
B
6
10
10
15
5
10
10
34.6
10.2
55
10
20
10
3

SECTION A-A
(SCALE 1:1)

R1
1
1
1
1

DETAIL B
(SCALE 2:1)

SHELL THICKNESS = 1MM
ALL INSIDE WALL THICKNESS

60°
60°
R46
R41
R50
R37
60°
9
60°
⌀12
⌀12
⌀12
60°
⌀12
10
⌀12
20

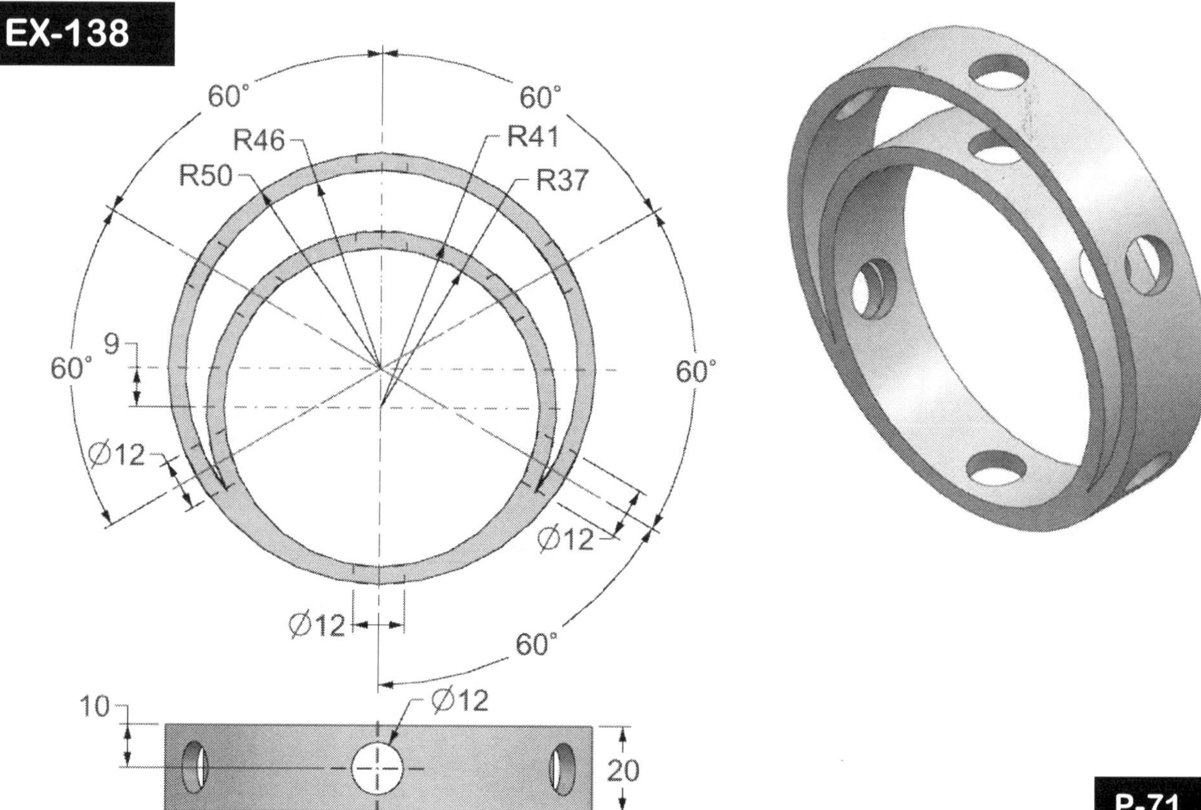

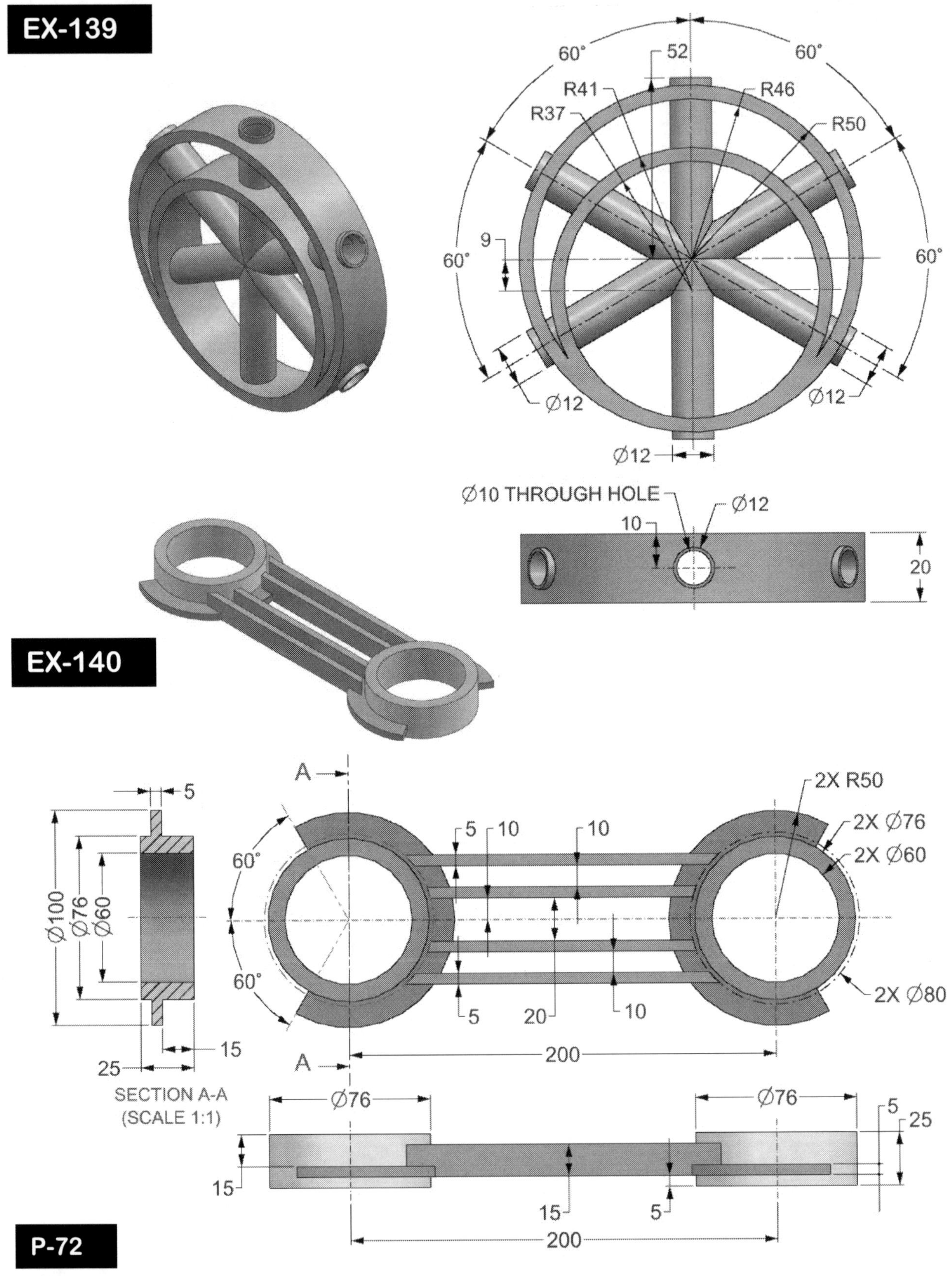

EX-139
60°
52
60°
R41
R46
R37
R50
9
60°
60°
60°
60°
Ø12
Ø12
Ø12
Ø10 THROUGH HOLE
Ø12
10
20
EX-140
A
5
2X R50
5
10
10
2X Ø76
60°
2X Ø60
Ø100
Ø76
Ø60
60°
5
20
10
2X Ø80
15
5
A
200
25
SECTION A-A
(SCALE 1:1)
Ø76
Ø76
5
25
15
15
5
200
P-72

EX-141 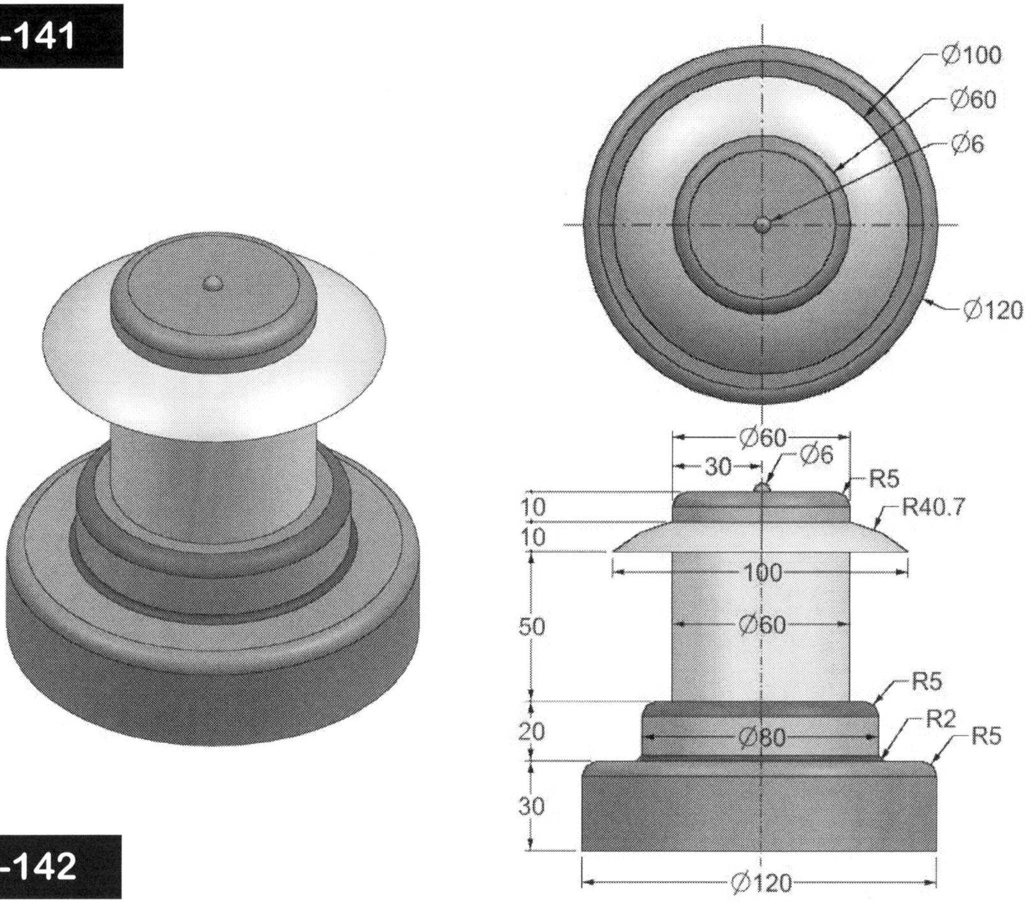
EX-142
P-73
Ø100
Ø60
Ø6
Ø120
Ø60
Ø6
R5
30
R40.7
10
10
100
50
Ø60
R5
20
Ø80
R2
R5
30
Ø120
18
14
15
41.4
15.4
R15
R10
46.8
Ø16.2
Ø8
8
23.4
Ø20.4
26
Ø16.2
R6
10 20
15
5

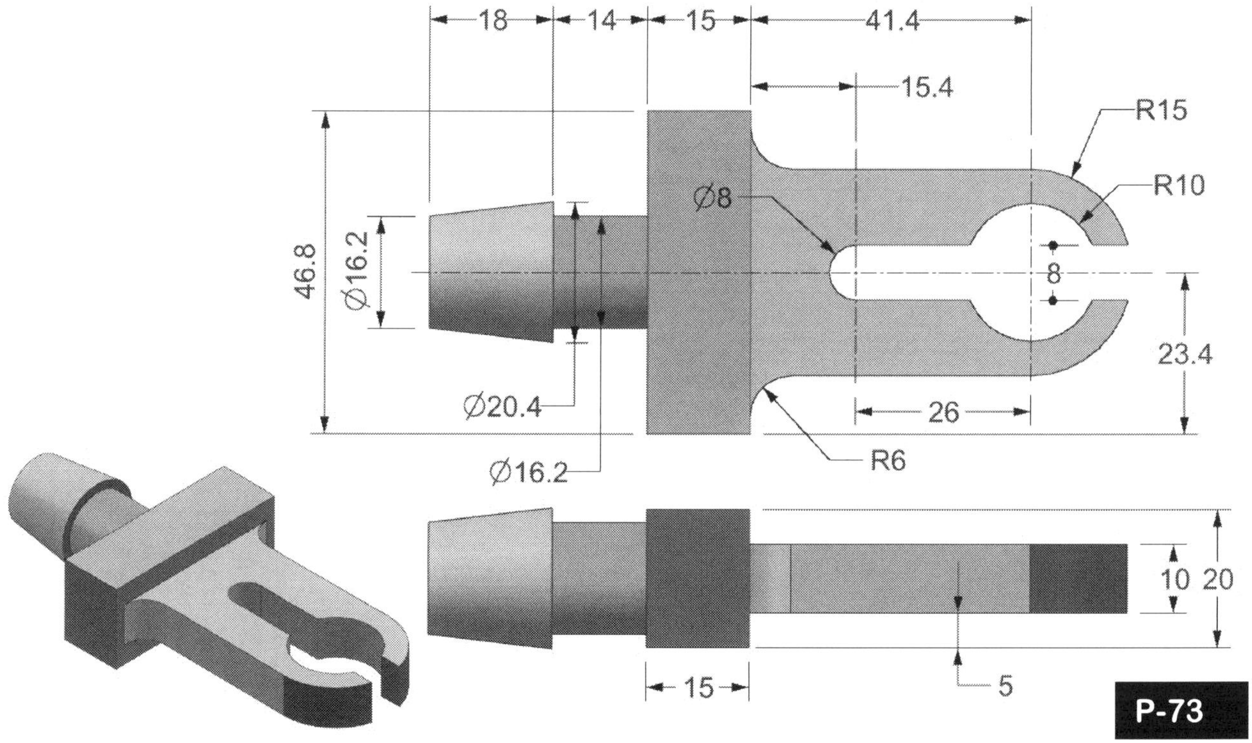

EX-143

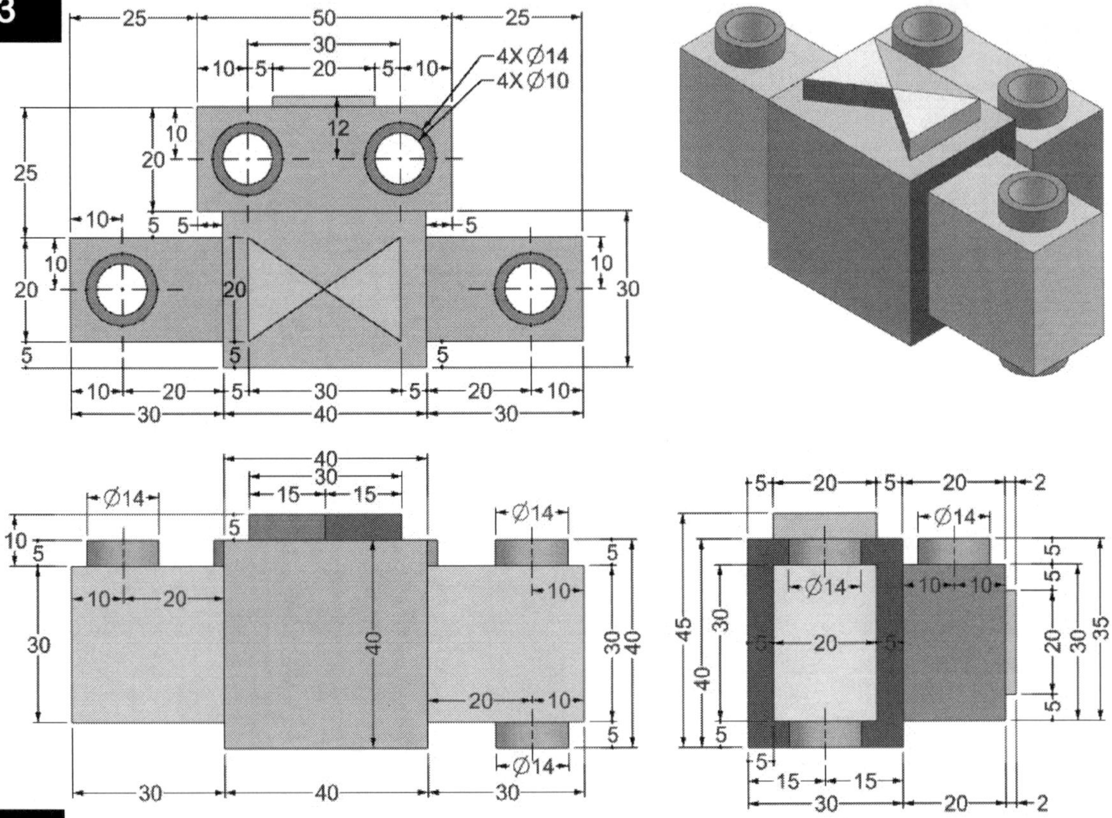
4X Ø14
4X Ø10

EX-144

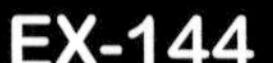

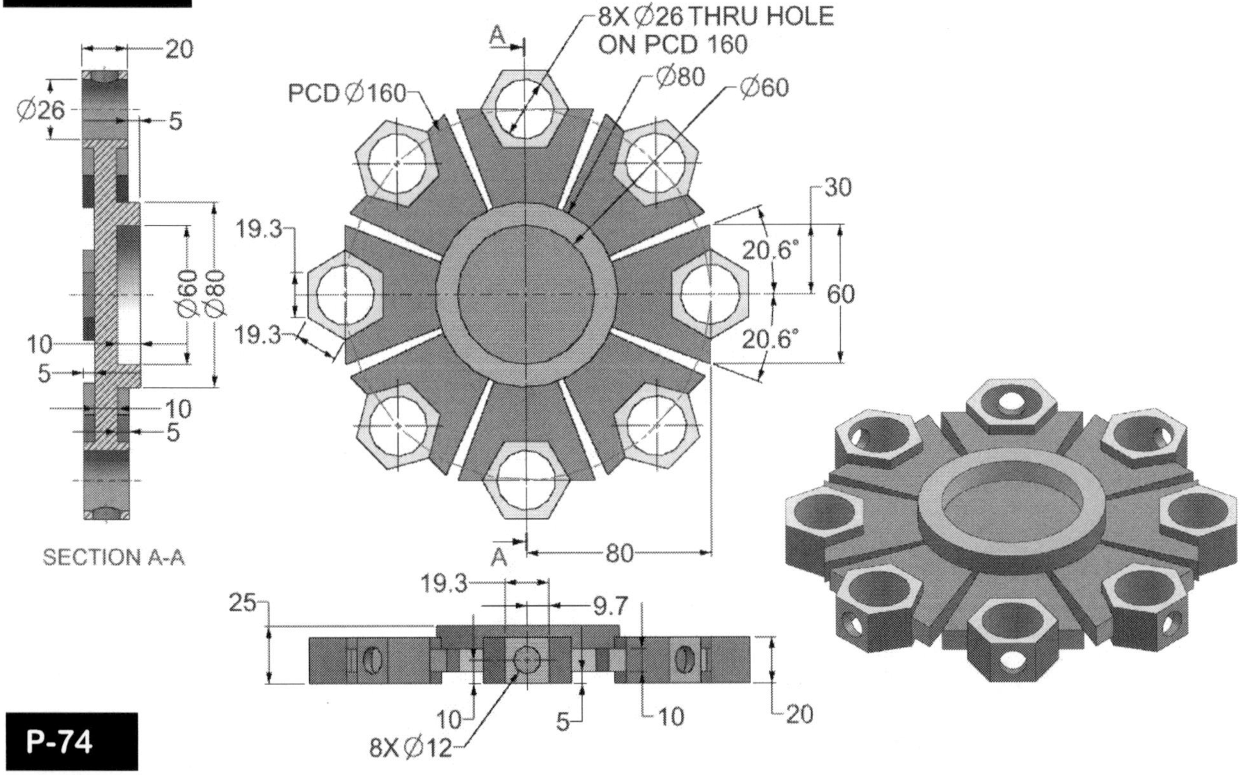
8X Ø26 THRU HOLE
ON PCD 160
PCD Ø160
Ø80
Ø60
20.6°
8X Ø12
SECTION A-A

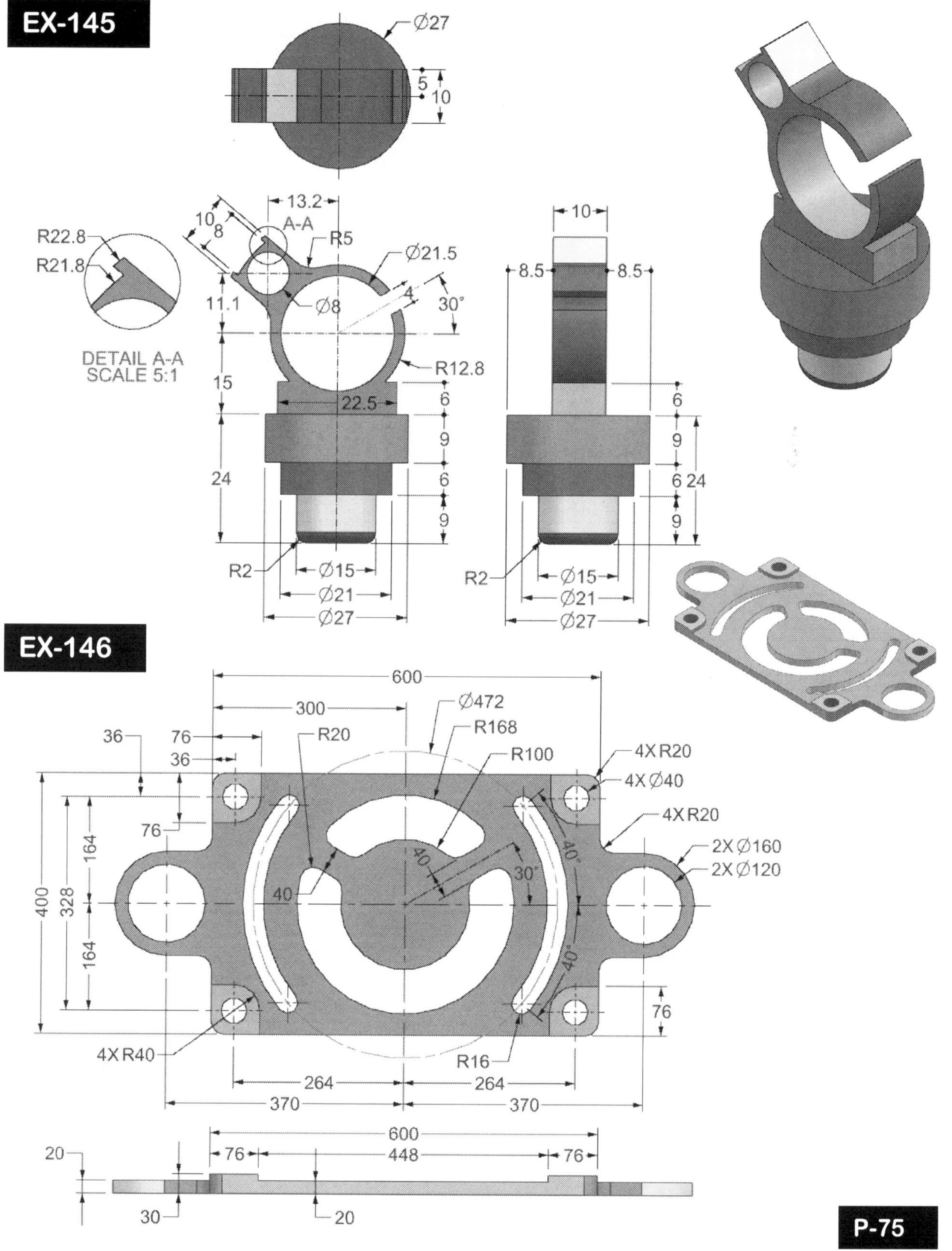

EX-145
Ø27
5
10
13.2
10
8
A-A
R5
Ø21.5
R22.8
R21.8
11.1
Ø8
4
30°
DETAIL A-A
SCALE 5:1
15
R12.8
22.5
6
9
24
6
9
R2
Ø15
Ø21
Ø27
10
8.5
8.5
6
9
6 24
9
R2
Ø15
Ø21
Ø27
EX-146
600
300
Ø472
R168
R20
R100
4X R20
4X Ø40
36
76
36
4X R20
76
2X Ø160
2X Ø120
164
40
30°
40°
400
328
40
40°
164
76
4X R40
R16
264
264
370
370
600
20
76
448
76
30
20
P-75

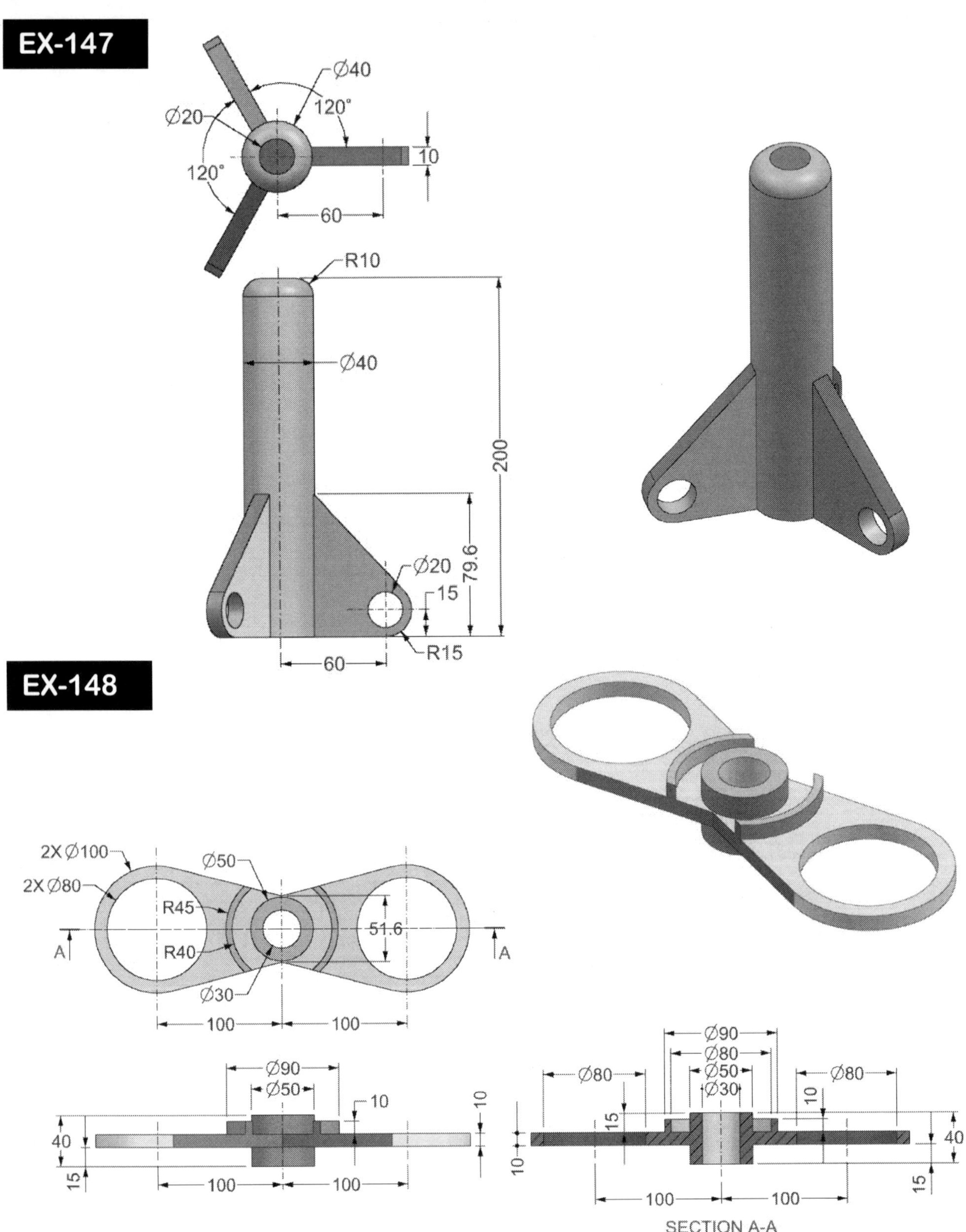

EX-147
Ø40
120°
Ø20
120°
10
60
R10
Ø40
200
79.6
Ø20
15
60
R15

EX-148
2X Ø100
Ø50
2X Ø80
R45
51.6
R40
A
A
Ø30
100
100
Ø90
Ø50
10
10
40
15
100
100
Ø80
Ø90
Ø80
Ø50
Ø30
Ø80
15
10
15
40
100
100
SECTION A-A

P-76

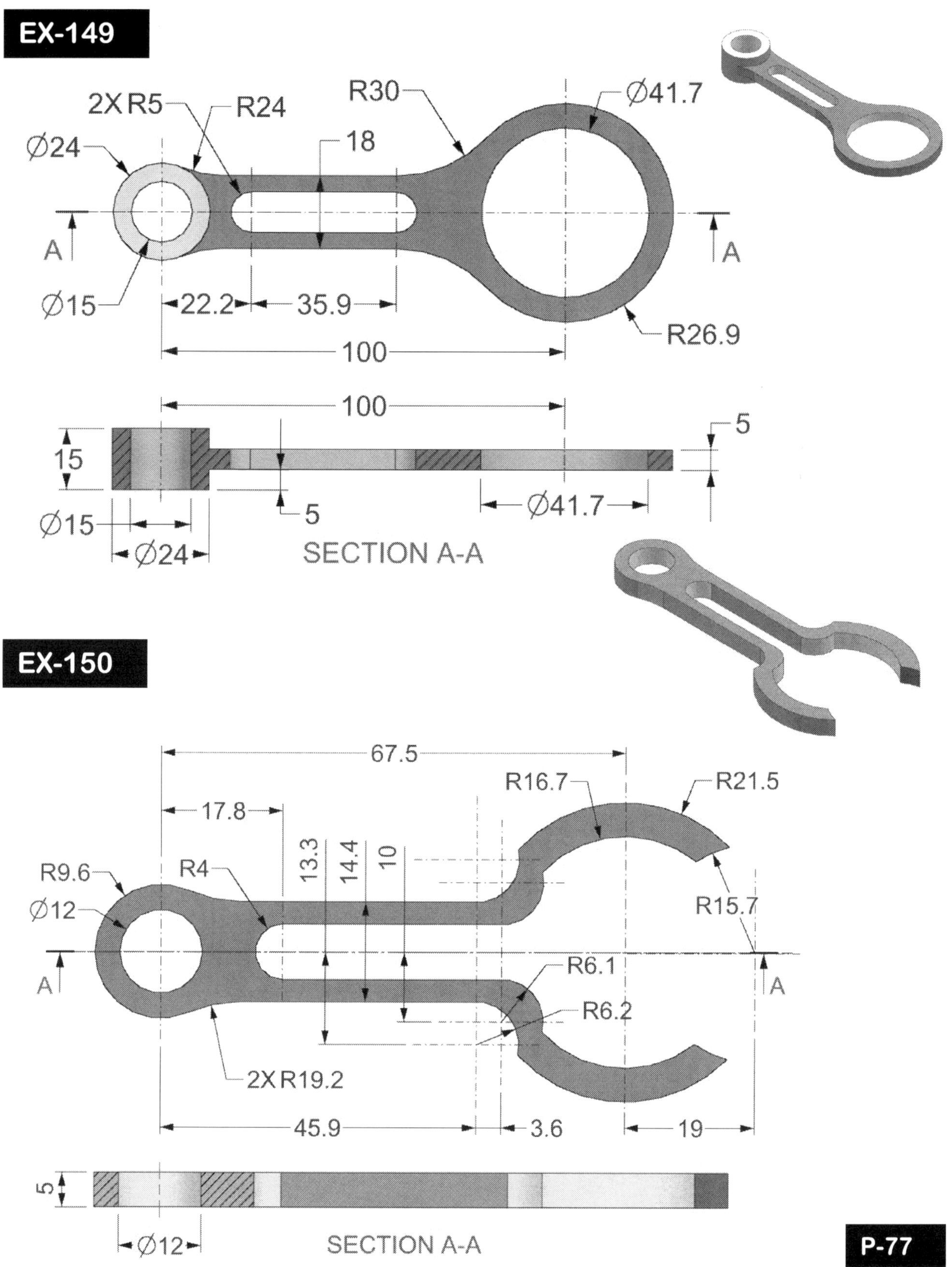

EX-149
Ø24
2X R5
R24
R30
18
Ø41.7
Ø15
A
A
22.2
35.9
100
R26.9
100
15
Ø15
Ø24
5
5
Ø41.7
SECTION A-A
EX-150
67.5
17.8
R16.7
R21.5
13.3
14.4
10
R9.6
R4
Ø12
R15.7
A
A
R6.1
R6.2
2X R19.2
45.9
3.6
19
5
Ø12
SECTION A-A
P-77

EX-151
Ø8 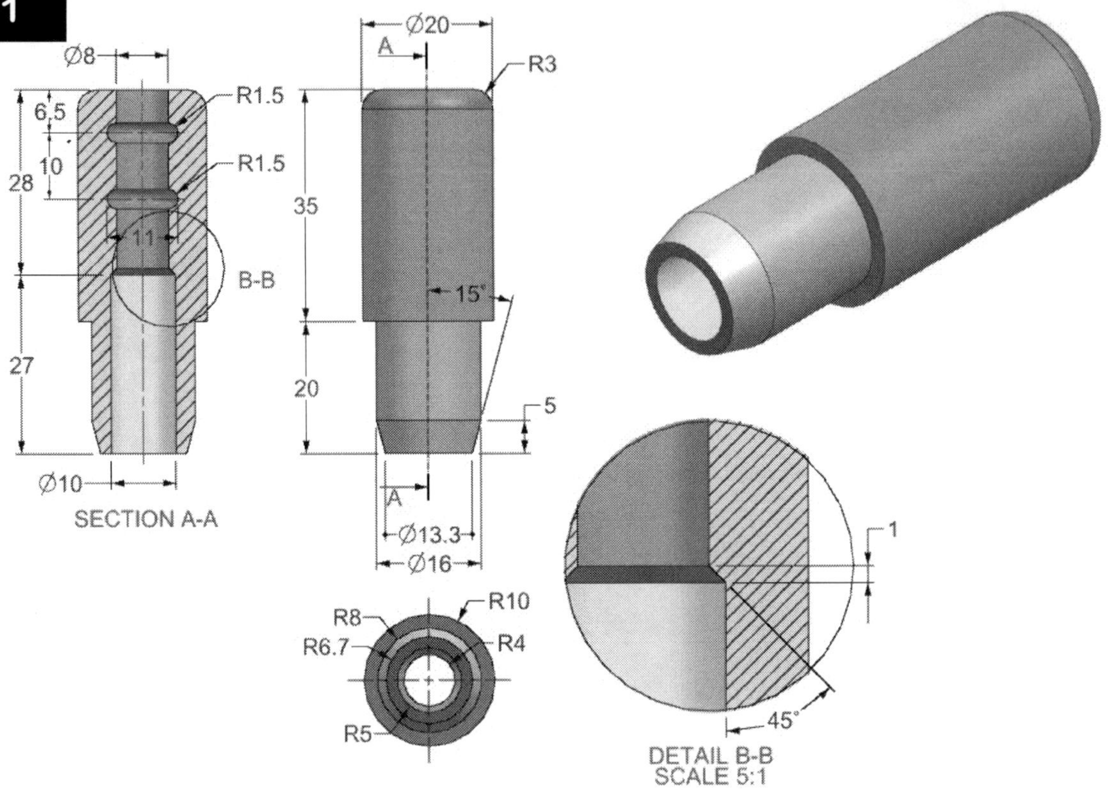
6.5
R1.5
10
28
R1.5
11
27
B-B
Ø10
SECTION A-A
Ø20
A
R3
35
15°
20
5
A
Ø13.3
Ø16
R8
R10
R6.7
R4
R5
1
45°
DETAIL B-B
SCALE 5:1
EX-152
Ø16 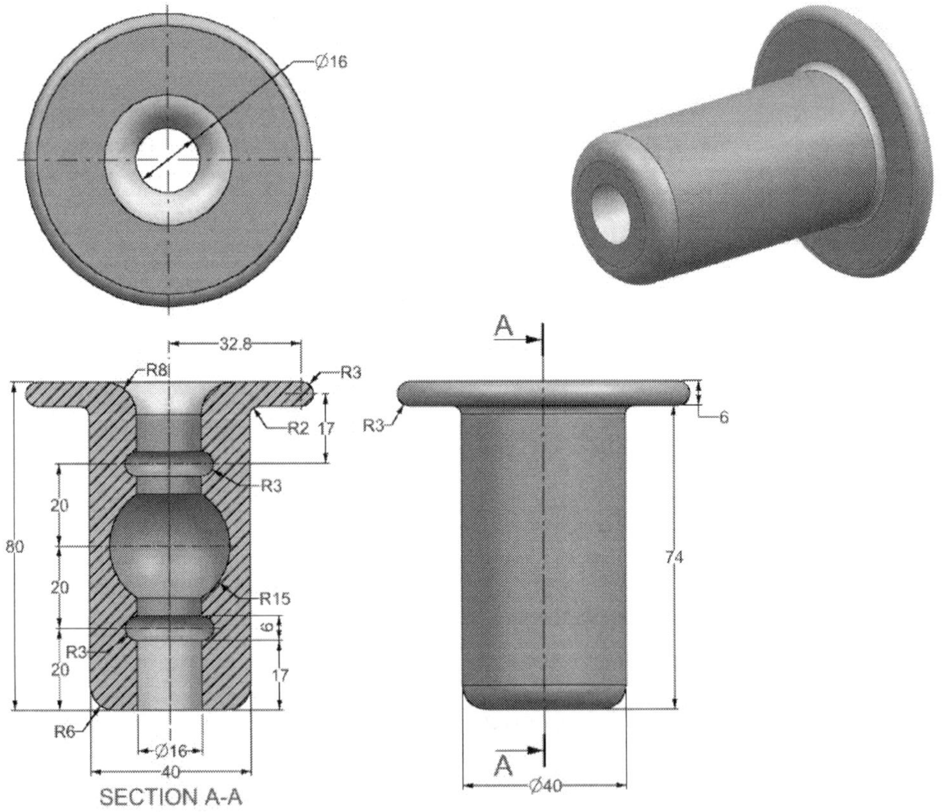
32.8
R8
R3
R2 17
20
R3
80
20
R15
6
R3
20
R6
Ø16
40
SECTION A-A
A
R3
6
74
A
Ø40
P-78

EX-153
EX-154
P-79
Ø16
Ø13.2
12.8°
30°
2.4
10.4
32
8.8
Ø12
2.4
Ø16
8
0.7
R2
6.4
4
14.4
30°
10.4
1.2
Ø10.4
1.2 X 45°
Ø12.8
Ø40
Ø12
Ø12
20
5.6
A
13.4
Ø40
13.4
48.4
R5
Ø24
13.7
35
R2.5
Ø20
9.9
R2.5
87.6°
Ø24
48.4
Ø8
11.4
SECTION A-A
A

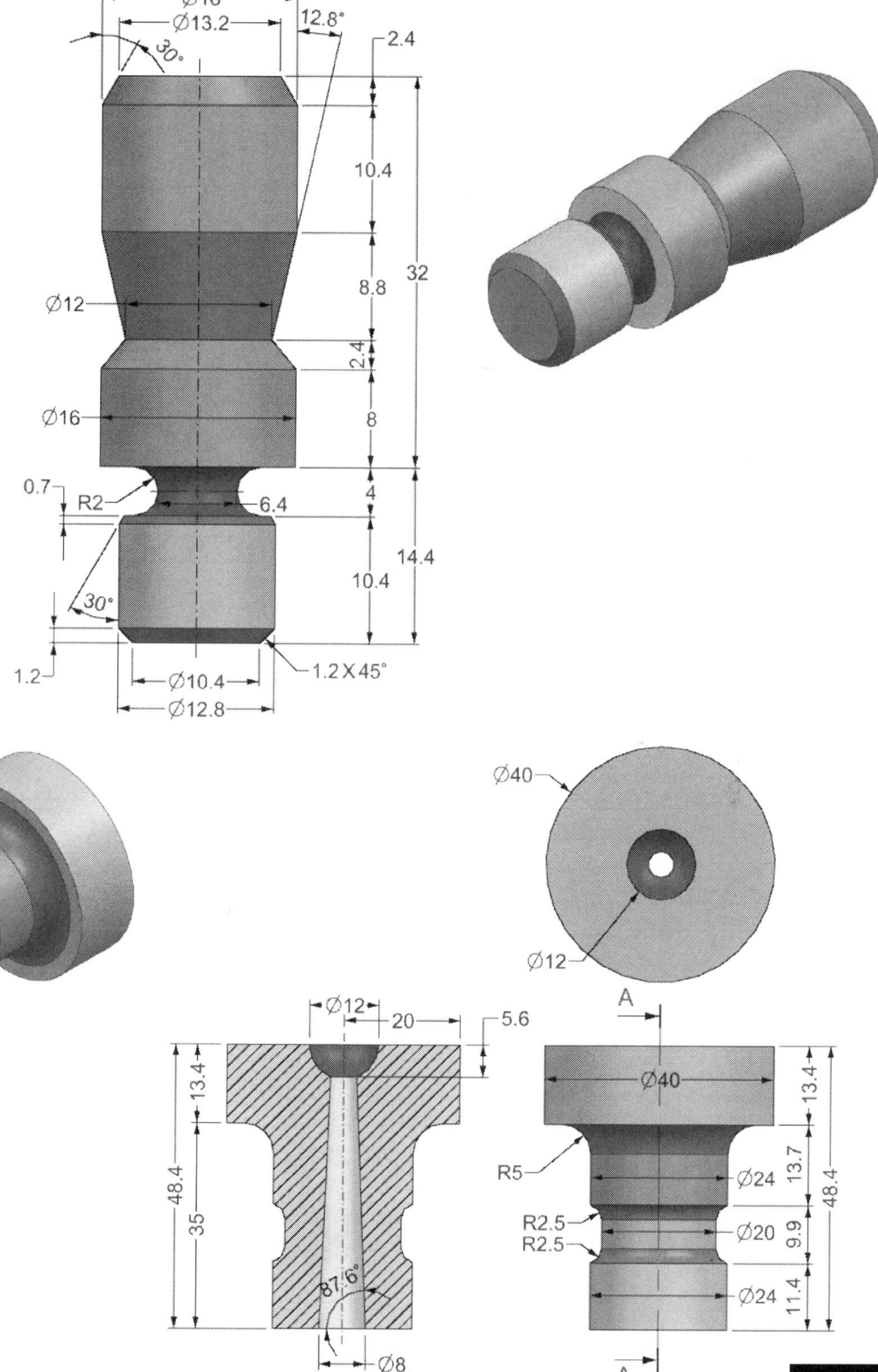

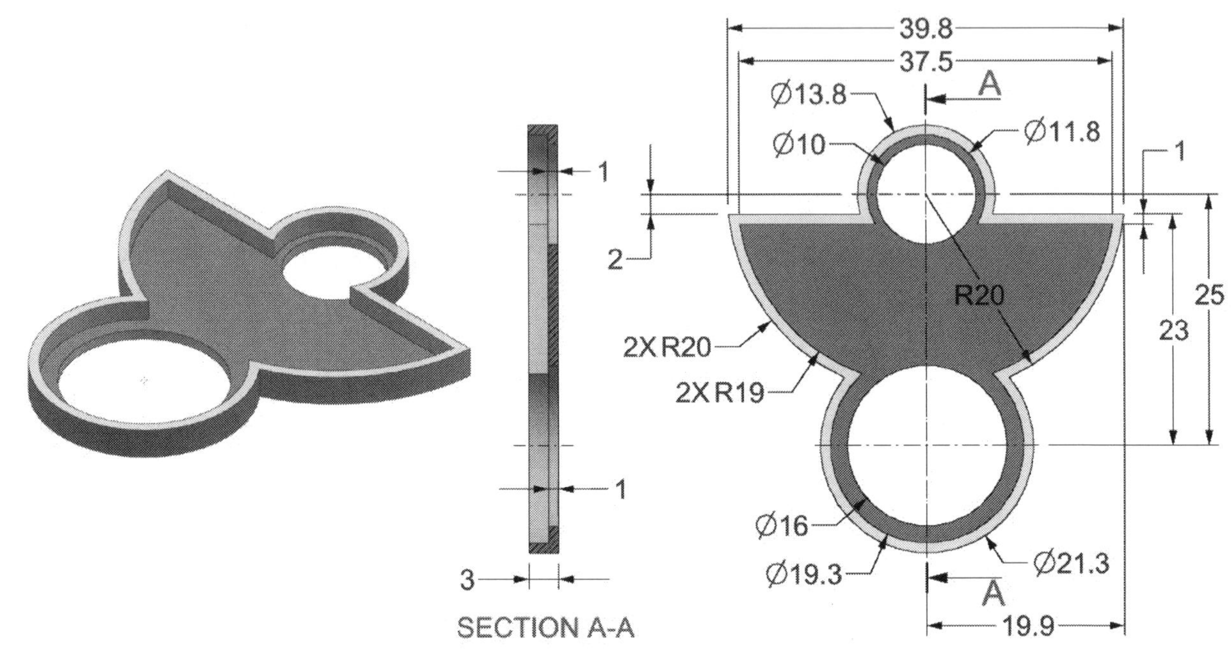

EX-155

39.8
37.5
A
Ø13.8
Ø11.8
Ø10
1
2
R20
25
23
2X R20
2X R19
1
Ø16
Ø19.3
Ø21.3
A
19.9
3

SECTION A-A

EX-156

Ø20
Ø14
30
Ø10.1
Ø27.4
5
10
A
A
22.8
Ø27.4
Ø10.1
3
134.8°
135°
SECTION A-A
Ø20
Ø14
30
10
Ø27.4
3
29.9
10
8.3
15.2
135°
42.86
14.28
57.14
15.2
10

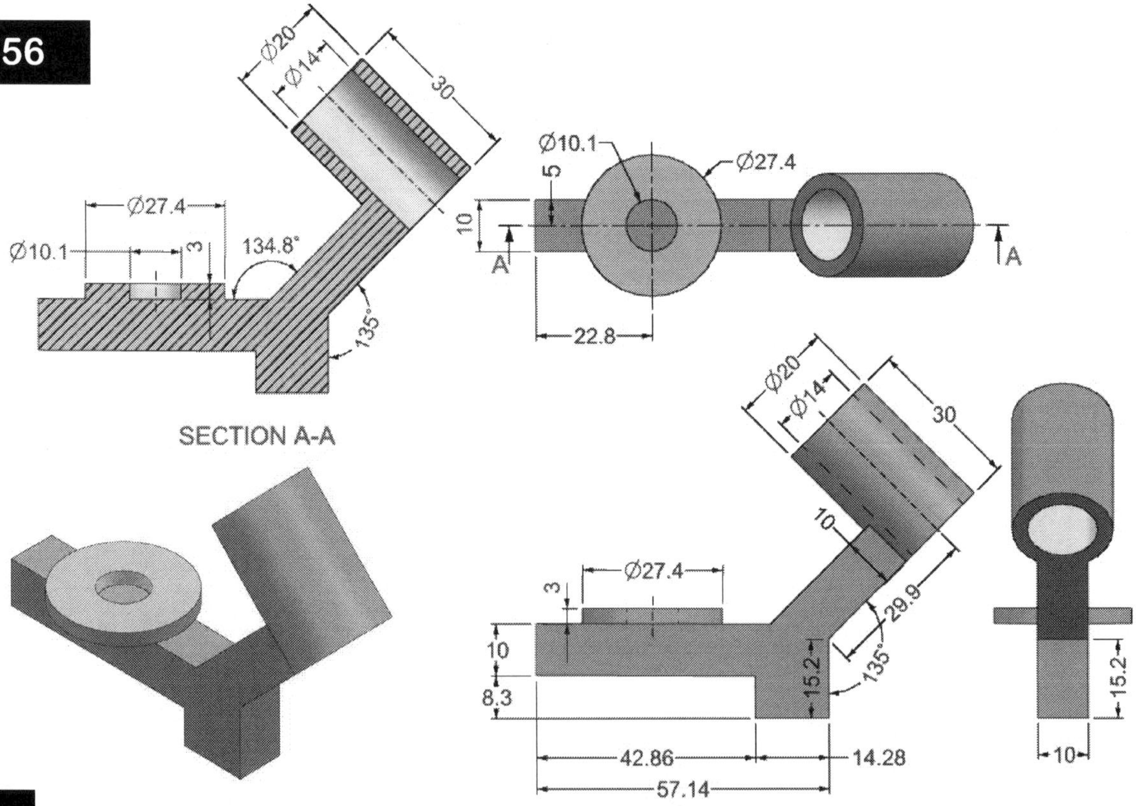

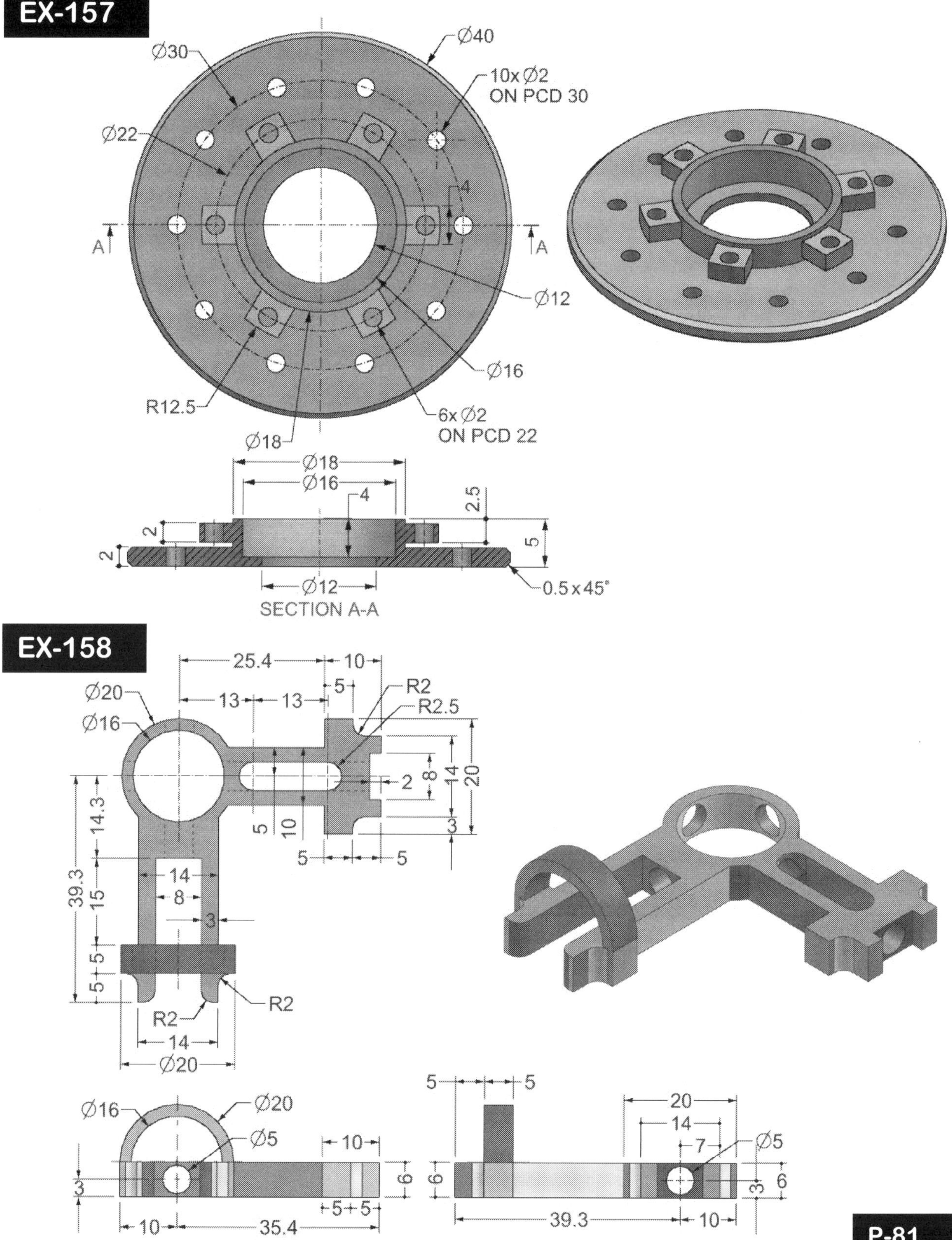

EX-157
Ø30
Ø22
Ø40
10x Ø2
ON PCD 30
4
Ø12
Ø16
R12.5
6x Ø2
ON PCD 22
Ø18
Ø18
Ø16
4
2.5
2
2
5
Ø12
0.5 x 45°
SECTION A-A
A
A

EX-158
25.4
10
5
13
13
R2
R2.5
Ø20
Ø16
14.3
2
8
14
20
5
10
3
5
5
39.3
14
8
15
3
5
5
R2
R2
14
Ø20
Ø16
Ø20
Ø5
10
3
10
35.4
5+5
6
5
5
6
20
14
7
Ø5
3
6
39.3
10

EX-159

EX-160

P-82

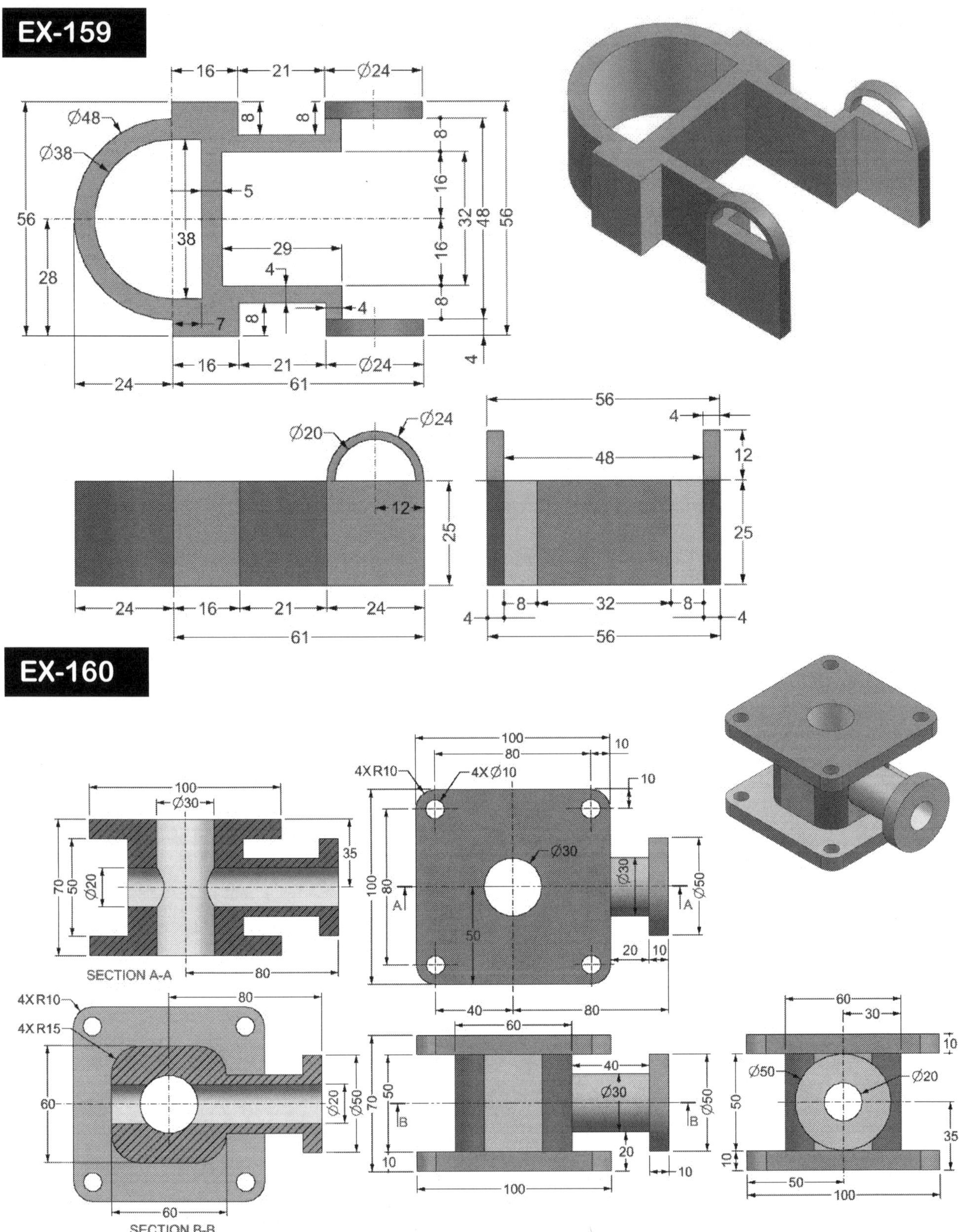

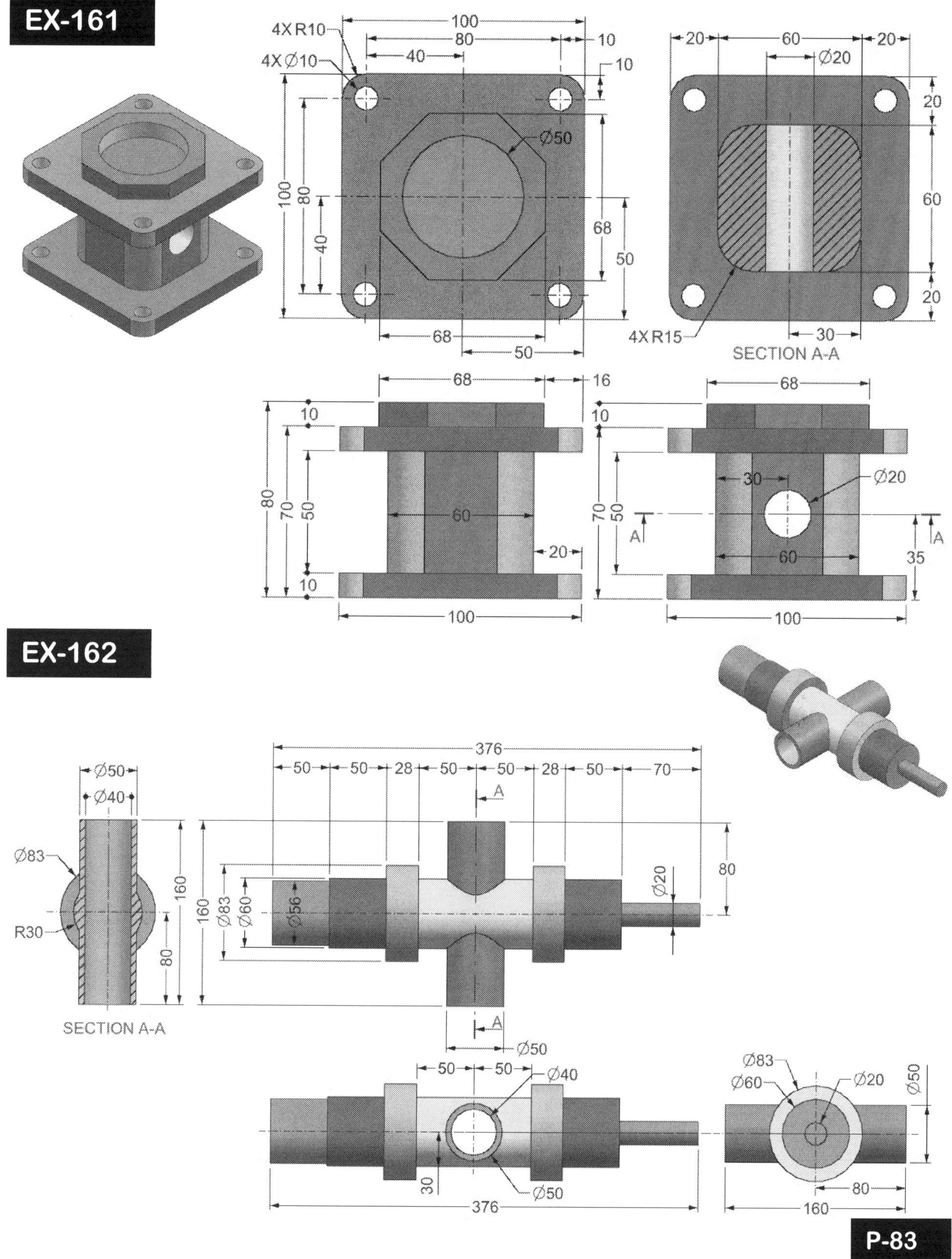

EX-161
4X R10
4X Ø10
100
80
40
10
10
Ø50
100
80
40
68
50
68
50
20
60
20
Ø20
20
60
20
4X R15
30
SECTION A-A
68
16
10
80
70
50
60
20
10
100
68
10
70
50
30
Ø20
A
60
A
35
100
EX-162
376
50
50
28
50
50
28
50
70
A
Ø50
Ø40
Ø83
R30
160
160
80
Ø83
Ø60
Ø56
80
Ø20
A
SECTION A-A
Ø50
50
50
Ø40
30
Ø50
376
Ø83
Ø60
Ø20
Ø50
80
160
P-83

EX-163

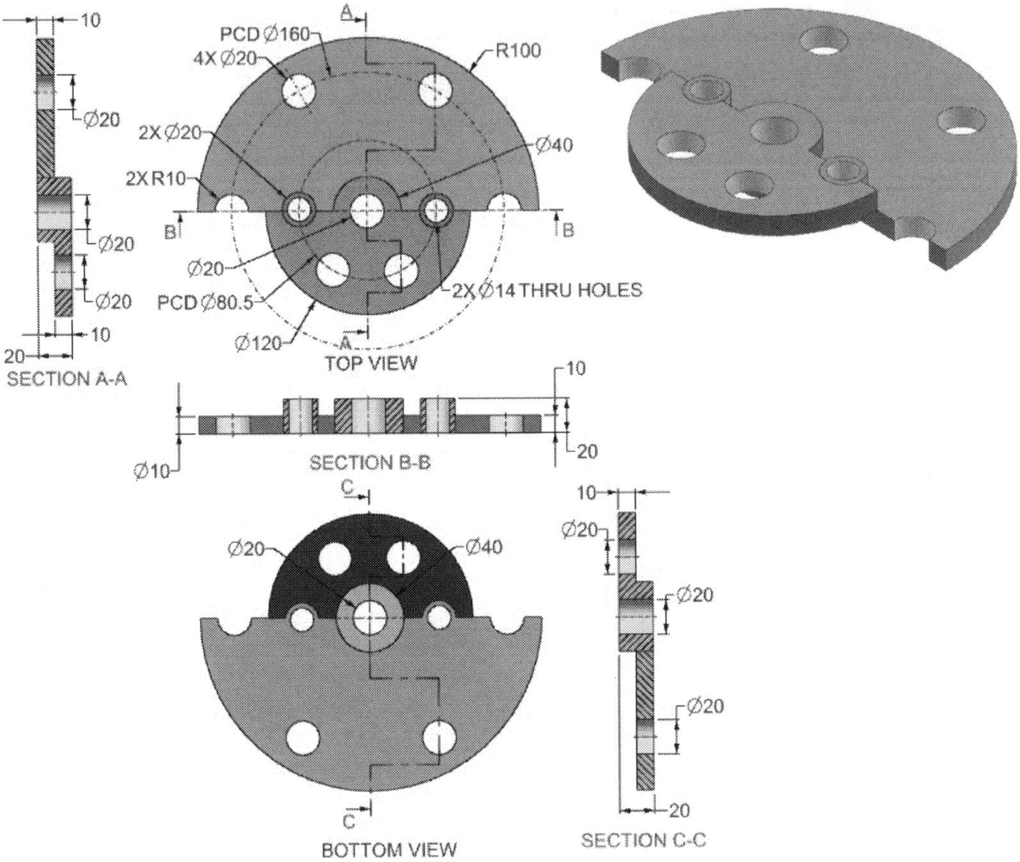

10
Ø20
Ø20
Ø20
20
10
Ø10
SECTION A-A

PCD Ø160
4X Ø20
2X Ø20
2X R10
Ø20
PCD Ø80.5
Ø120
A
R100
Ø40
B
B
2X Ø14 THRU HOLES
A
TOP VIEW

10
20
SECTION B-B

C
Ø20
Ø40
Ø20
C
BOTTOM VIEW

10
Ø20
Ø20
Ø20
20
SECTION C-C

EX-164

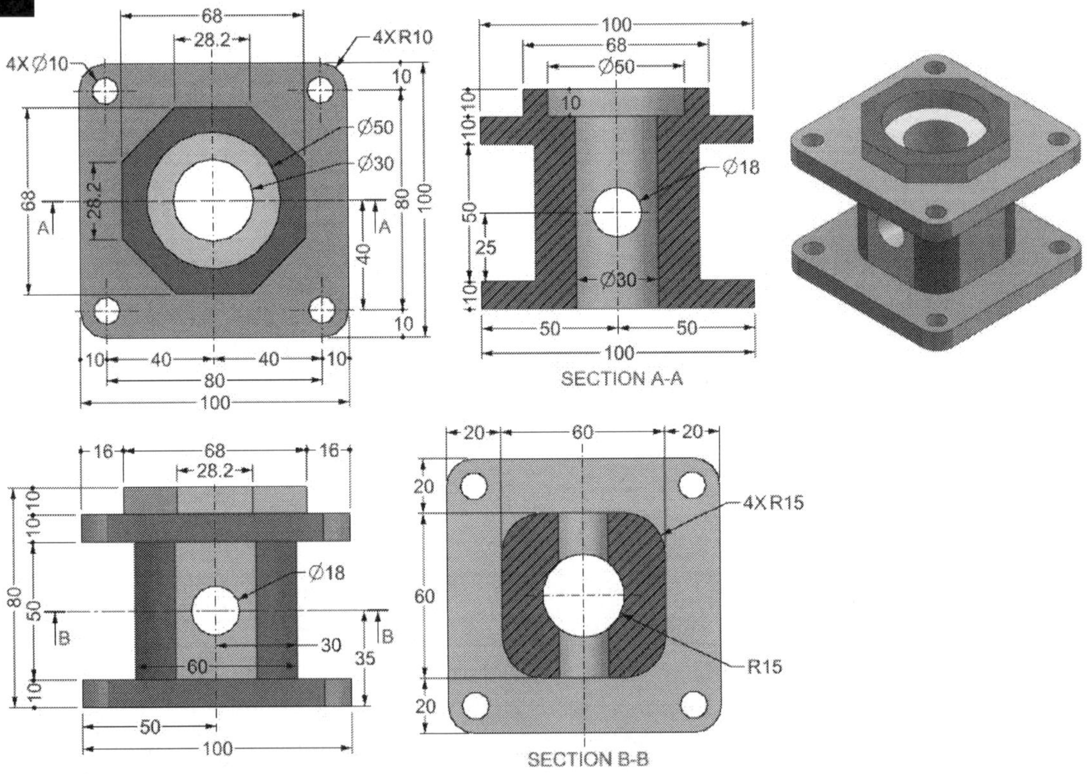

68
28.2
4X Ø10
4X R10
10
Ø50
Ø30
68
28.2
A
80
100
40
A
10
10
40
40
10
80
100

100
68
Ø50
10
10 10
Ø18
50
25
10
Ø30
50
50
100
SECTION A-A

16
68
16
28.2
10 10
80
50
B
10
Ø18
60
30
35
B
50
100

20
60
20
20
4X R15
60
R15
20
SECTION B-B

P-84

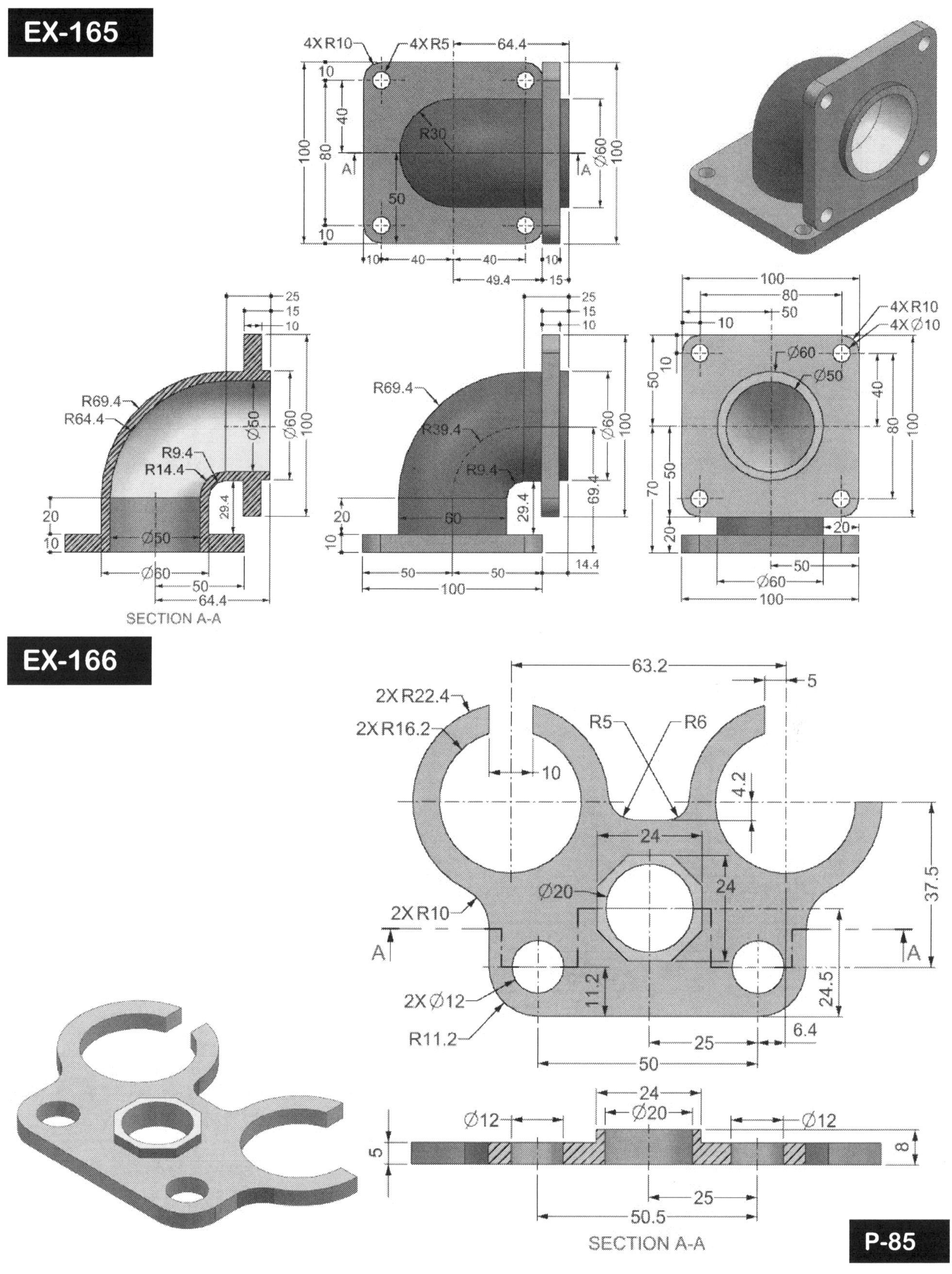
4X R10
4X R5
64.4
10
40
100
80
R30
A
50
10
Ø60
100
A
10
40
40
10
49.4
15
25
15
10
R69.4
R64.4
Ø50
Ø60
100
R9.4
R14.4
29.4
20
10
Ø50
Ø60
50
64.4
SECTION A-A
25
15
10
R69.4
Ø60
100
R39.4
R9.4
69.4
29.4
20
10
60
50
50
14.4
100
100
80
50
10
4X R10
4X Ø10
Ø60
50
Ø50
40
80
100
70
20
20
SECTION A-A
50
50
Ø60
100
2X R22.4
2X R16.2
63.2
5
R5
R6
10
4.2
24
24
Ø20
2X R10
A
37.5
A
11.2
24.5
2X Ø12
6.4
R11.2
25
50
24
Ø12
Ø20
Ø12
5
8
25
50.5
SECTION A-A
P-85

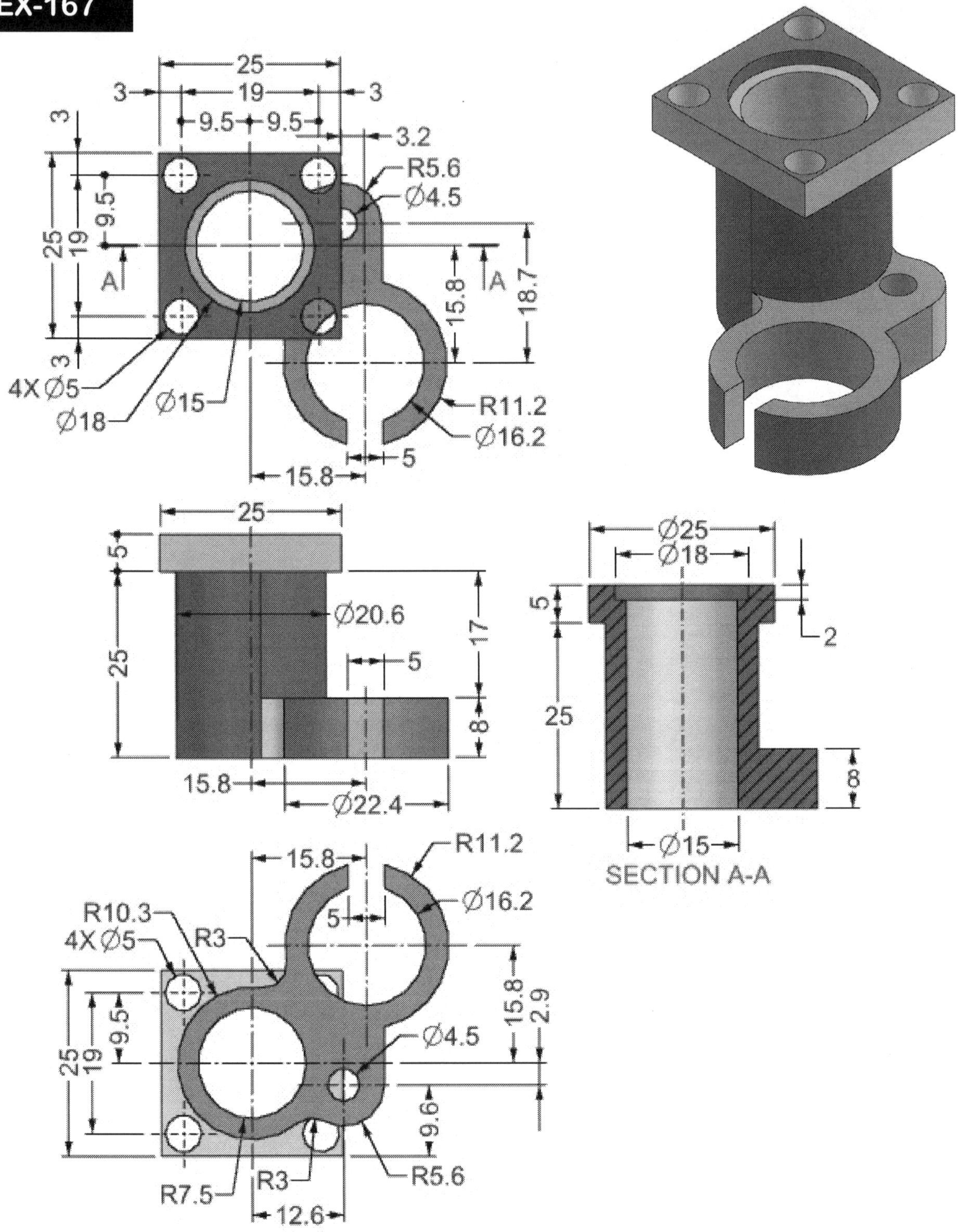

25
3
19
3
9.5
9.5
3.2
R5.6
Ø4.5
3
9.5
25
19
A
15.8
A
18.7
4X Ø5
Ø15
Ø18
R11.2
Ø16.2
3
5
15.8

25
5
Ø20.6
25
17
5
8
15.8
Ø22.4

Ø25
Ø18
5
2
25
8
Ø15
SECTION A-A

15.8
R11.2
R10.3
5
Ø16.2
4X Ø5
R3
15.8
2.9
Ø4.5
25
19
9.5
9.6
R7.5
R3
R5.6
12.6

EX-168 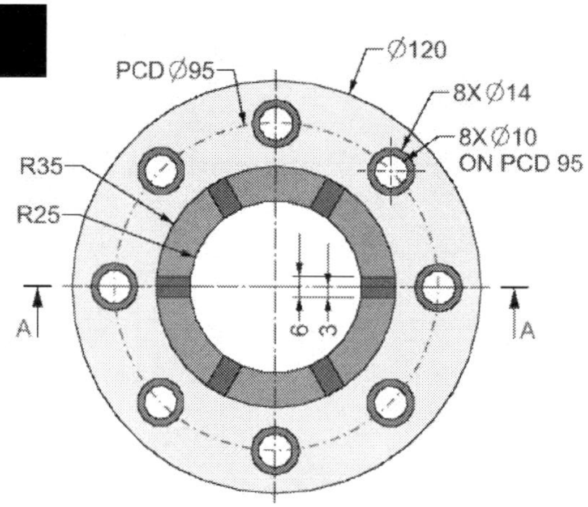 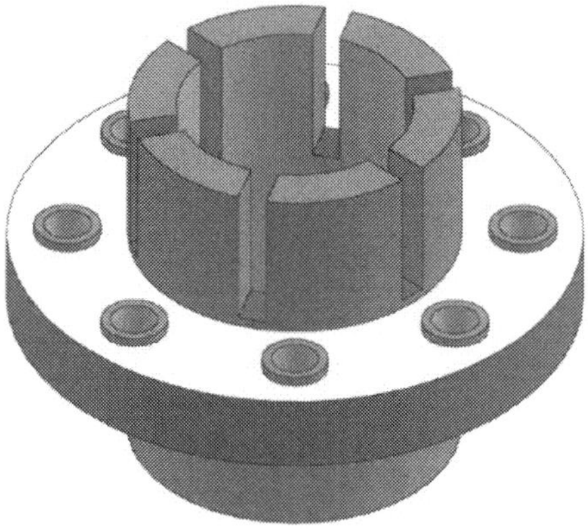 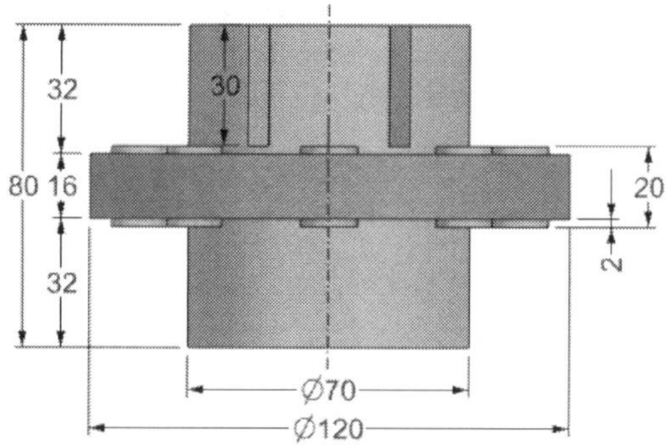 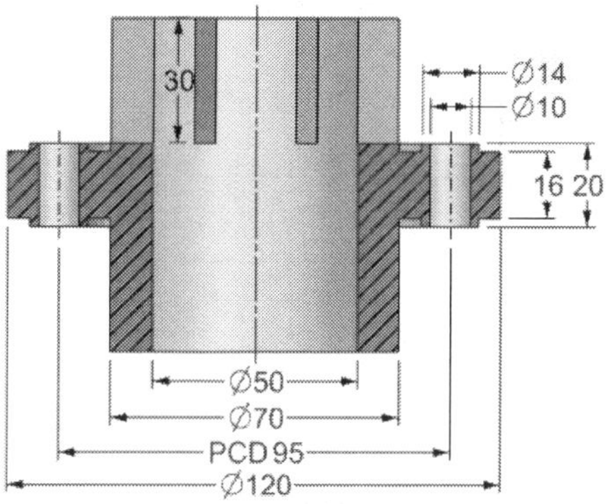
PCD ∅95
∅120
8X ∅14
8X ∅10
ON PCD 95
R35
R25
A
A
6
3
32
30
80 16
32
20
2
∅70
∅120
30
∅14
∅10
16 20
∅50
∅70
PCD 95
∅120
SECTION A-A

EX-169
∅70
∅40
20
R5
40
∅28
∅40
50
130
70
200
∅70
R2
R60
∅80
R5
30
40
21.3
50
140
80
10
∅28
∅40
50
80
70
∅70
15
40
15
10
30
∅28
∅40
80
15
15
70
30
10
35
∅70

P-87

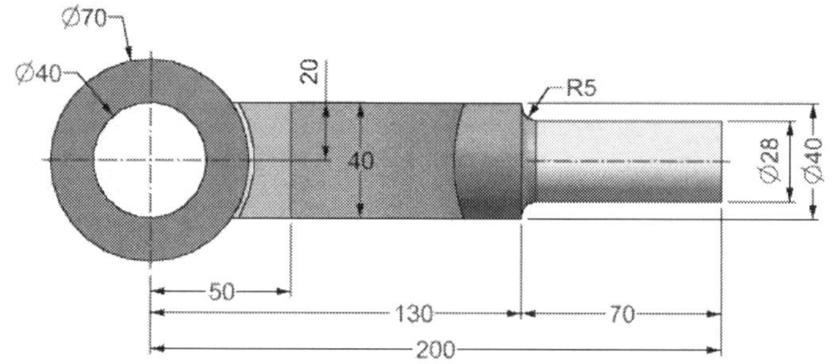

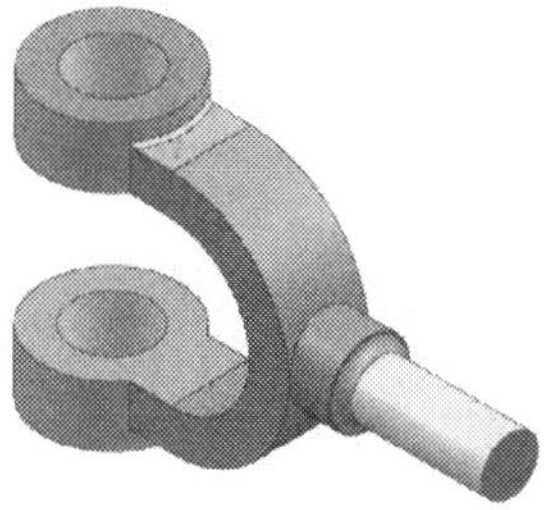

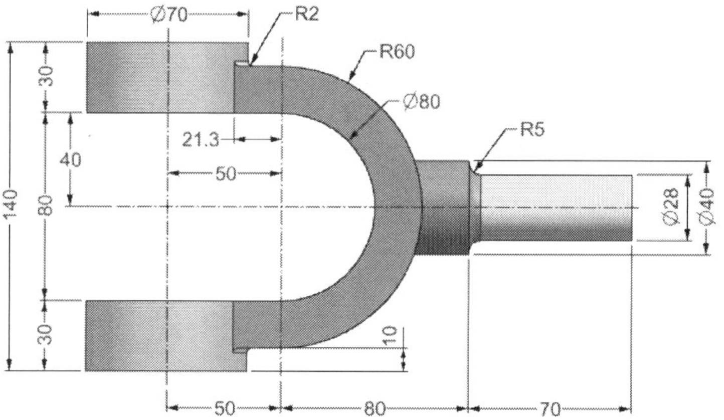

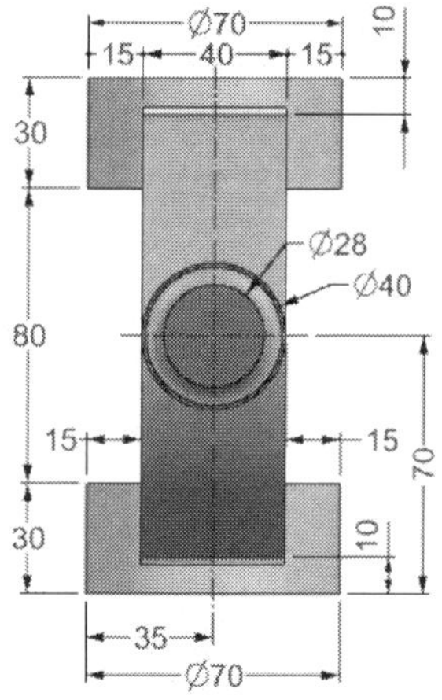

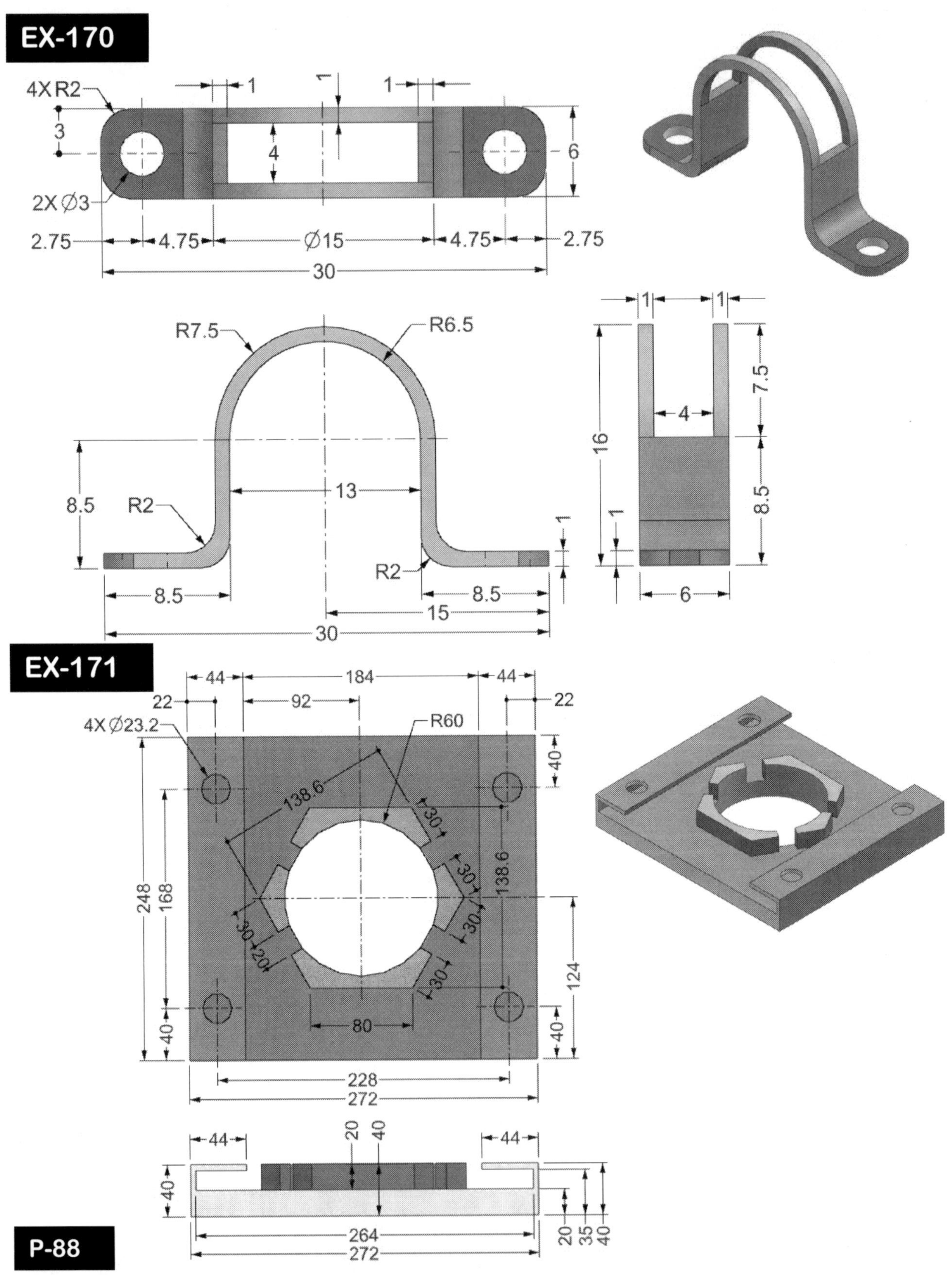

EX-170
4X R2
3
2X Ø3
2.75
4.75
Ø15
4.75
2.75
30
1
1
1
4
6
R7.5
R6.5
8.5
R2
13
R2
8.5
8.5
15
30
1
1
16
4
7.5
8.5
6
EX-171
44
184
44
22
22
4X Ø23.2
R60
138.6
30
30
30
30
30
138.6
248
168
20
80
40
124
40
228
272
44
20
40
44
40
264
272
20
35
40
P-88

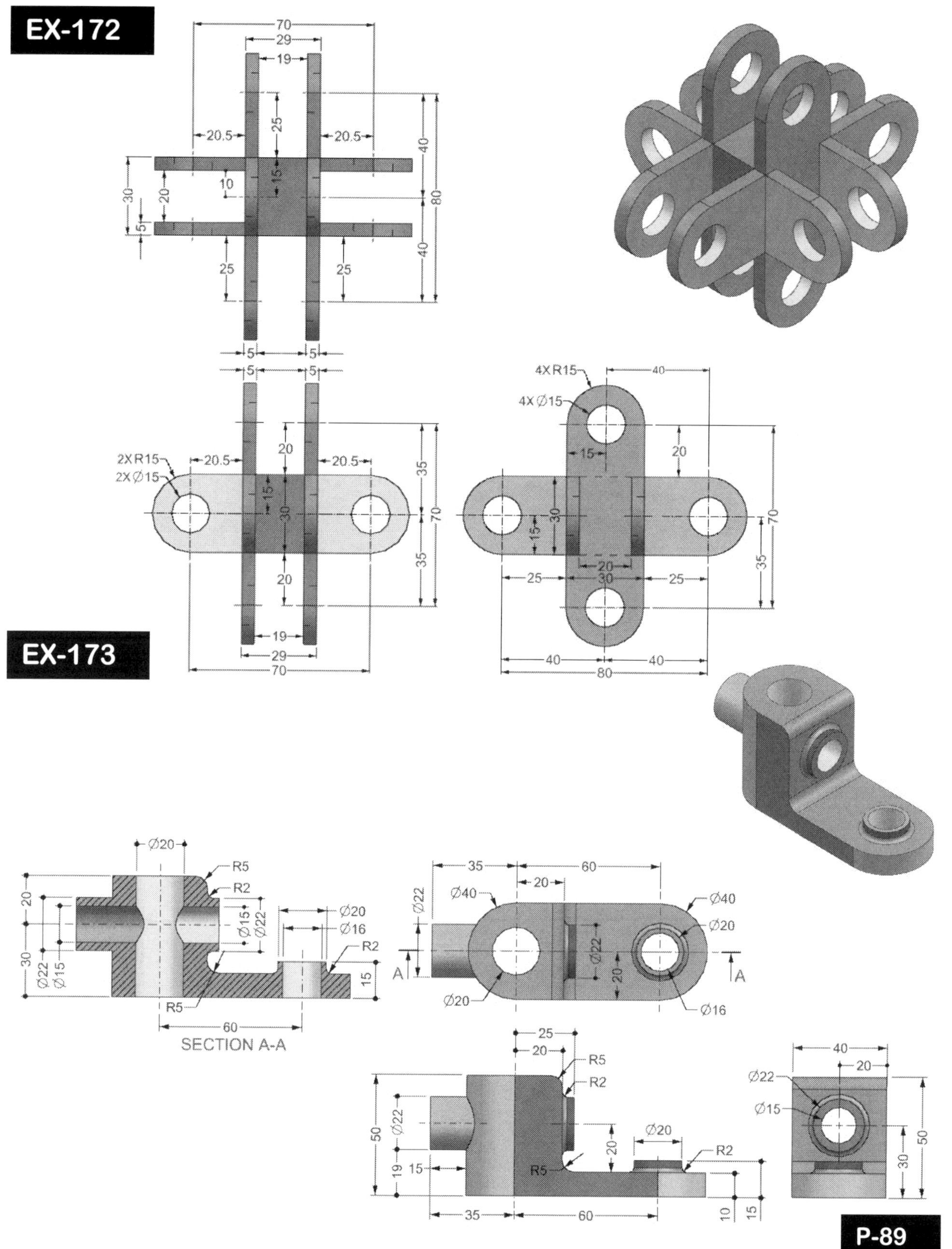

EX-172
EX-173
P-89
SECTION A-A
2X R15
2X Ø15
4X R15
4X Ø15
Ø20
Ø16
Ø22
Ø15
Ø40
R5
R2
A
A

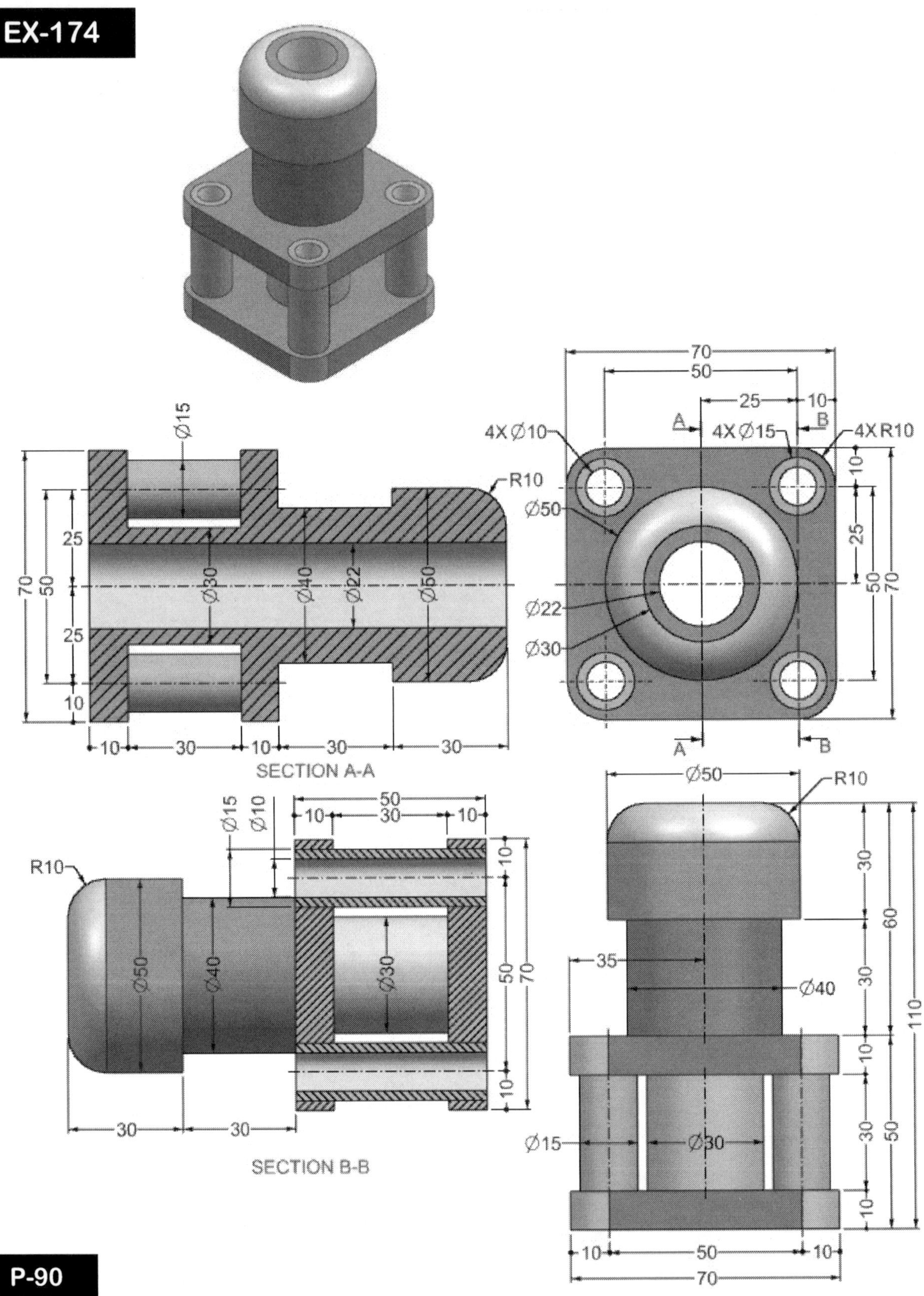

Ø15
R10
25
70
50
25
10
Ø30
Ø40
Ø22
Ø50
10
30
10
30
30
SECTION A-A
70
50
25
10
4X Ø10
Ø50
Ø22
Ø30
A
4X Ø15
B
4X R10
10
25
50
70
A
B
Ø15
Ø10
50
10
30
10
10
R10
Ø50
Ø40
Ø30
50
70
30
30
10
SECTION B-B
Ø50
R10
30
60
35
Ø40
30
110
10
50
30
Ø15
Ø30
10
10
50
10
70

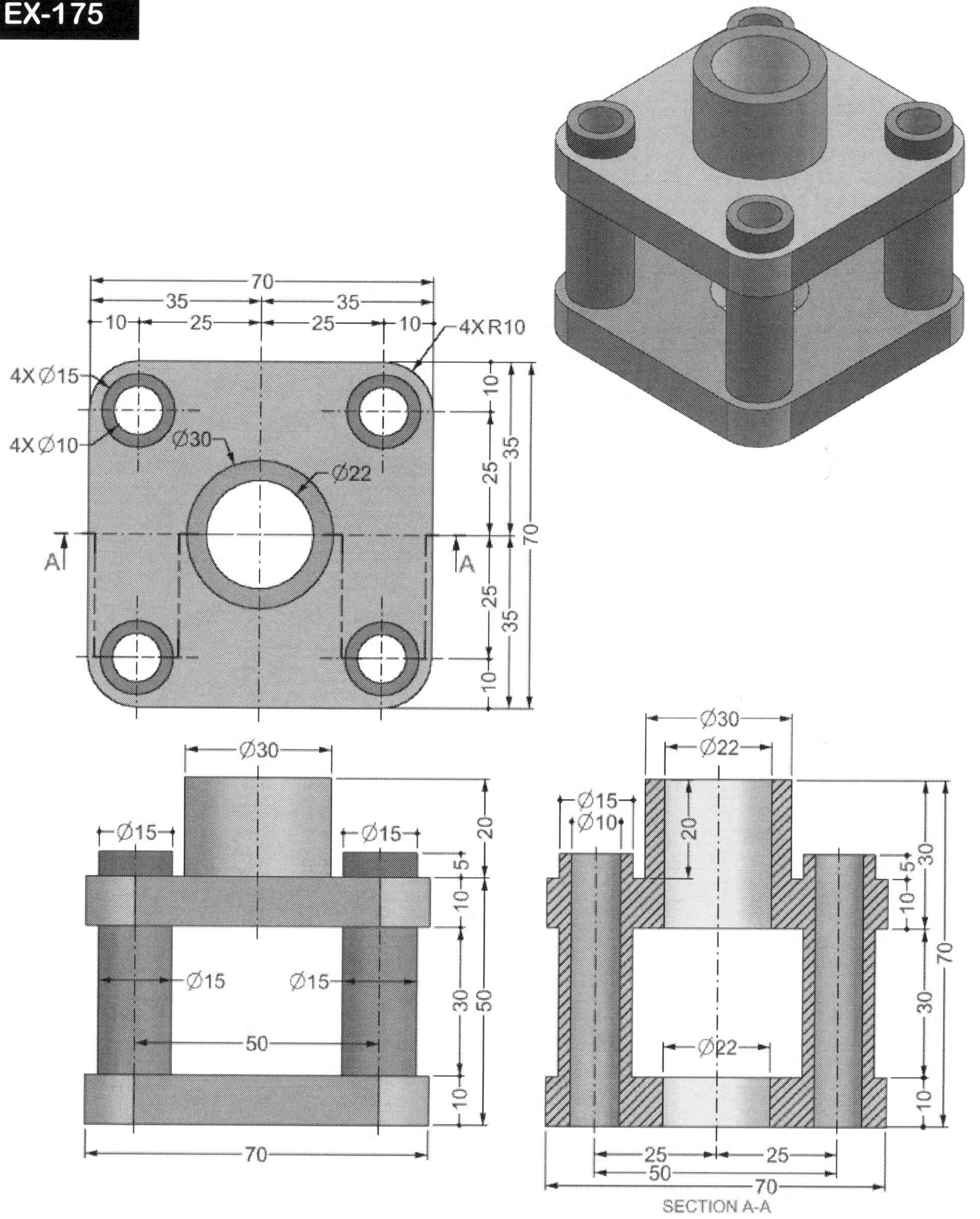

70
35 35
10 25 25 10 4X R10
4X Ø15
4X Ø10
Ø30
Ø22
10
35
25
70
A
A
25
35
10

Ø30
Ø15 Ø15
20
5
10
Ø15 Ø15
30
50
50
10
70

Ø30
Ø22
Ø15
Ø10
20
30
5
10
70
Ø22
30
10
25 25
50
70
SECTION A-A

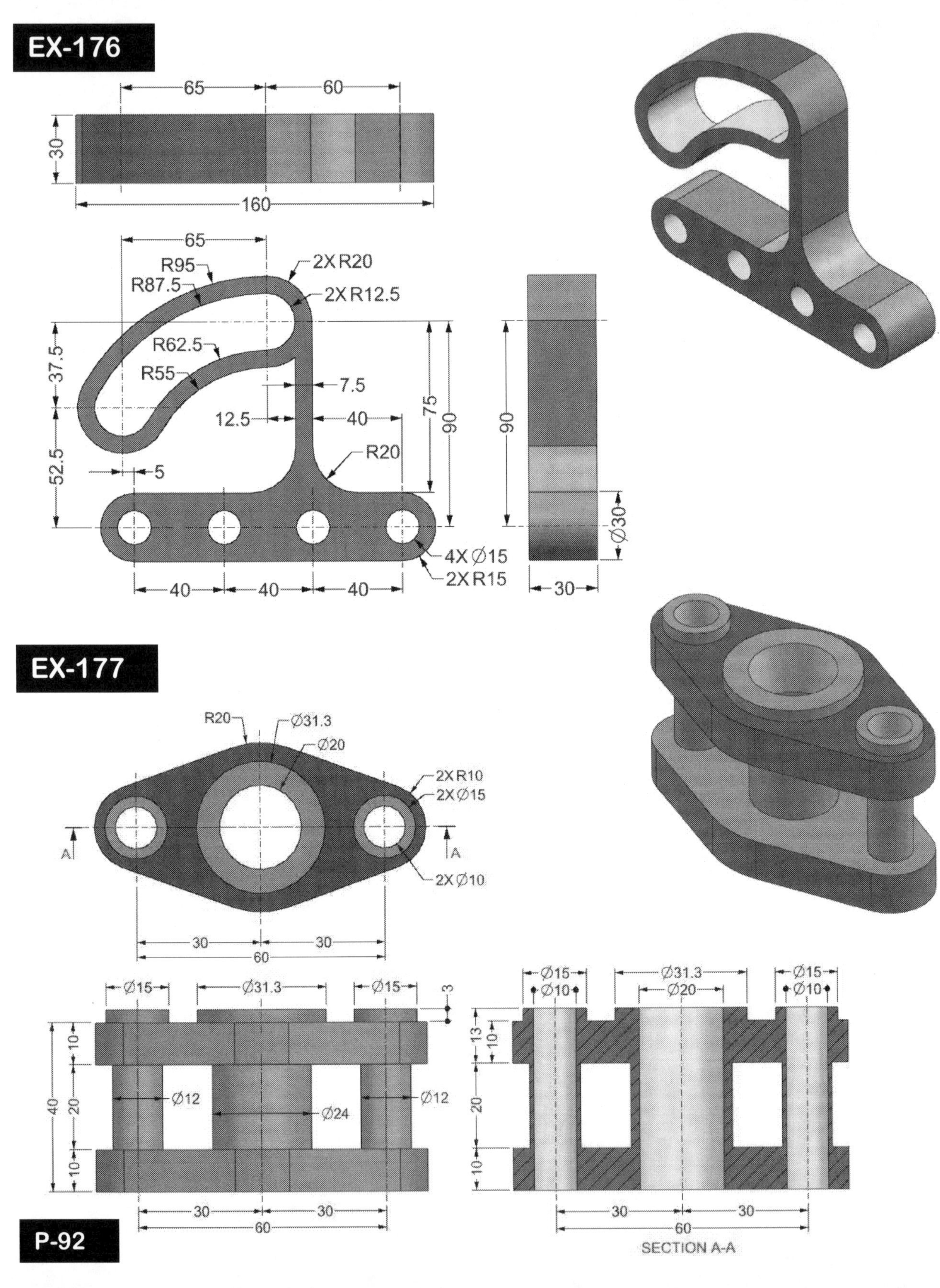

EX-176
65
60
30
160
65
R95
R87.5
2X R20
2X R12.5
R62.5
R55
7.5
37.5
12.5
40
75
90
90
52.5
5
R20
4X Ø15
2X R15
40
40
40
30
Ø30
EX-177
R20
Ø31.3
Ø20
2X R10
2X Ø15
A
A
2X Ø10
30
30
60
Ø15
Ø31.3
Ø15
3
Ø15
Ø10
Ø31.3
Ø20
Ø15
Ø10
10
13
10
40
20
Ø12
Ø24
Ø12
20
10
10
30
30
30
30
60
60
P-92
SECTION A-A

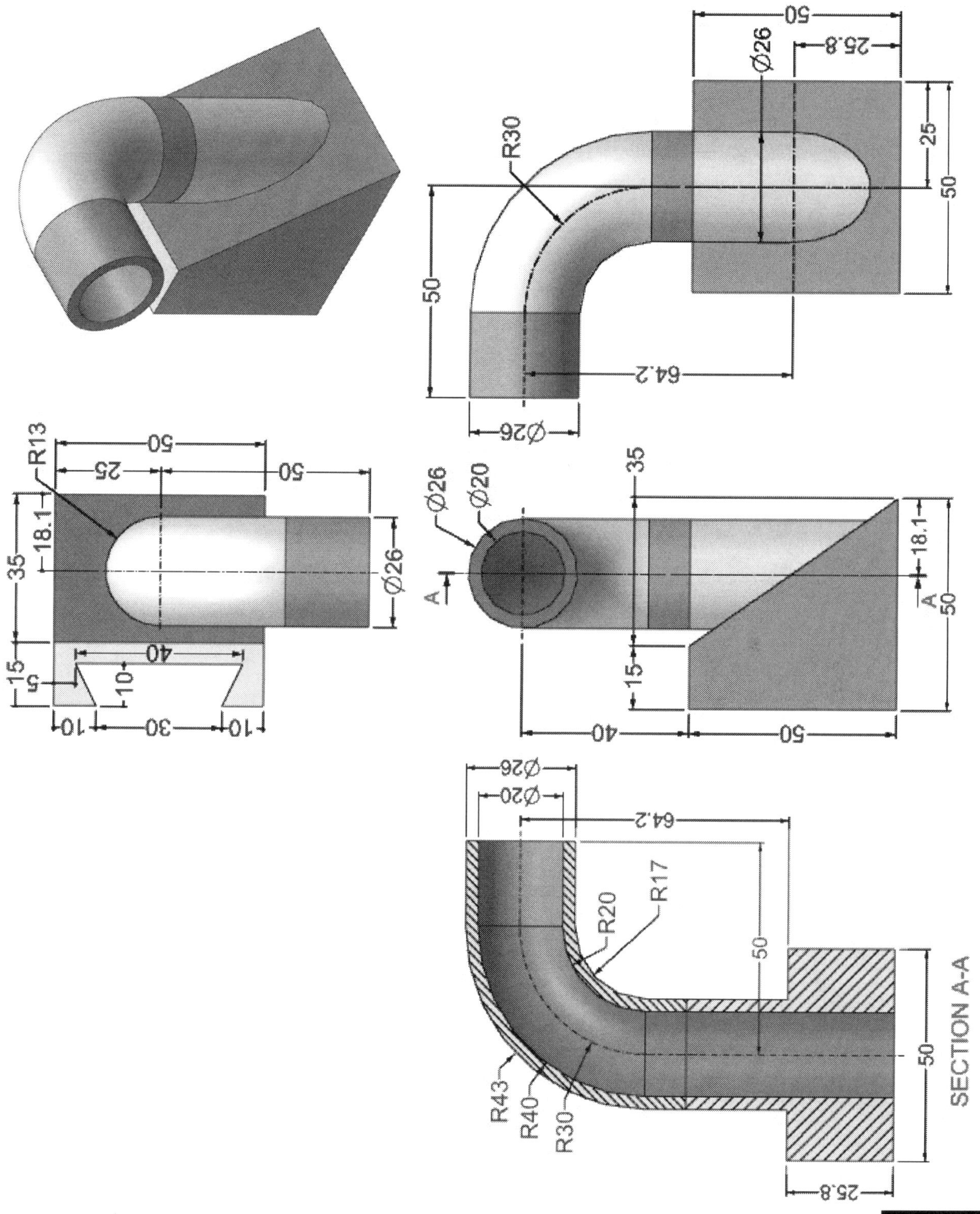

R30
Ø26
50
25.8
25
50
64.2
Ø26
R13
50
25
50
35
18.1
Ø26
40
15
5
10
30
10
Ø26
Ø20
35
A
A
15
40
50
18.1
50
Ø26
Ø20
64.2
R20
R17
50
R43
R40
R30
50
25.8
SECTION A-A

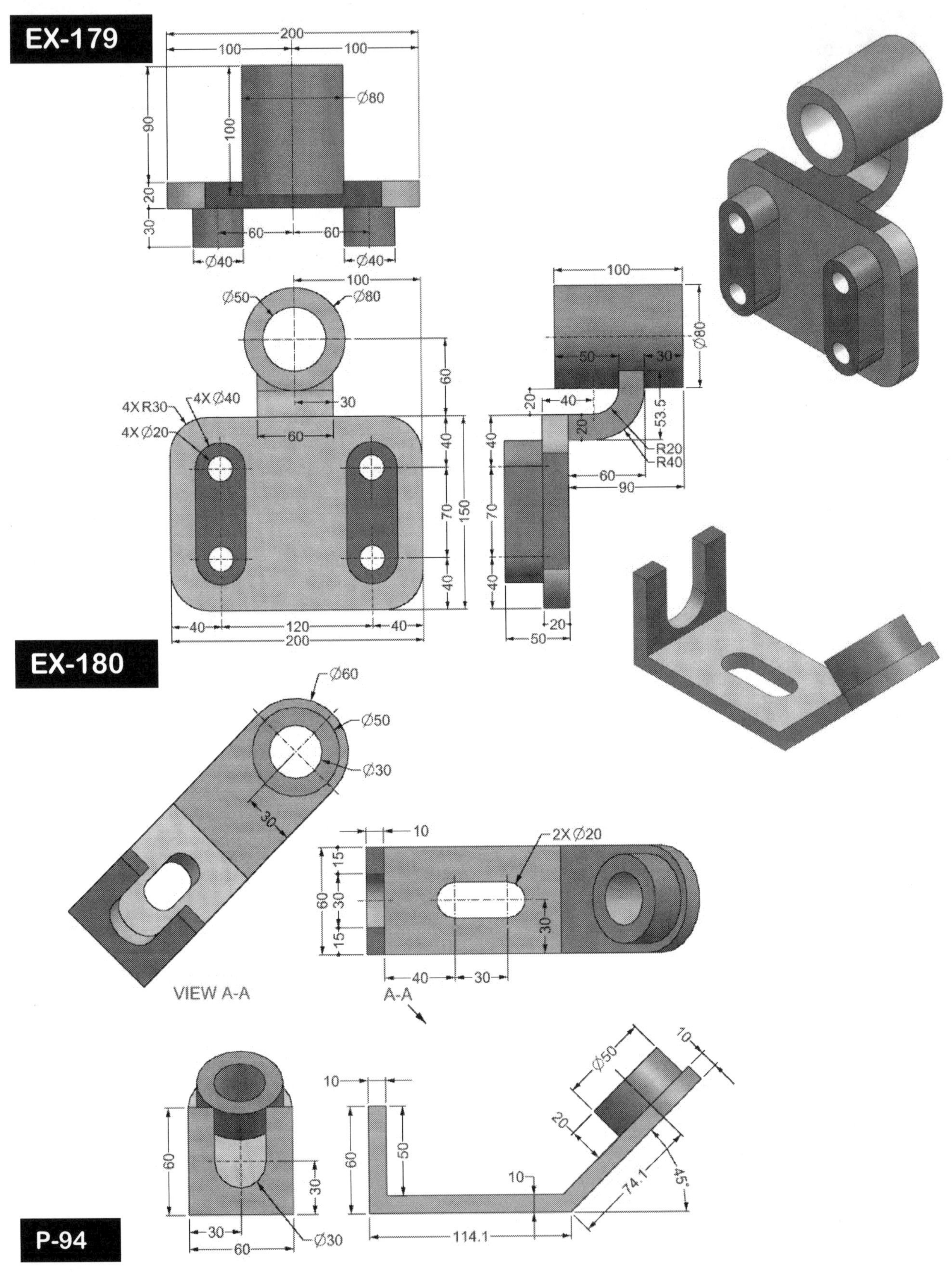

EX-179
EX-180
P-94
Ø80
200
100
100
90
100
20
30
60
60
Ø40
Ø40
Ø50
Ø80
100
4X R30
4X Ø40
4X Ø20
30
60
60
40
70
150
60
40
40
40
120
40
200
70
40
20
50
100
50
30
20
40
20
60
90
R20
R40
53.5
Ø80
Ø60
Ø50
Ø30
30
10
15
60
30
15
2X Ø20
30
40
30
VIEW A-A
A-A
60
30
30
60
Ø30
10
10
60
50
114.1
Ø50
10
20
74.1
45°

EX-181 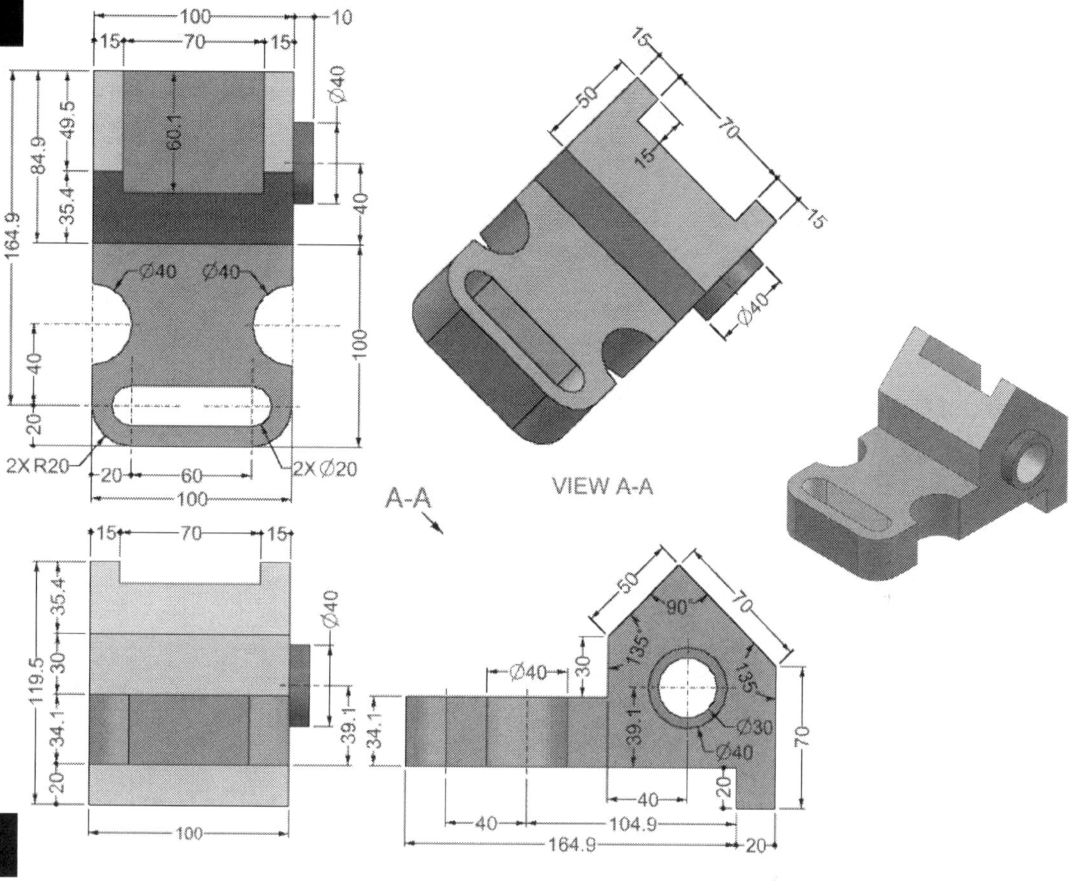
100
10
15
70
15
Ø40
60.1
49.5
84.9
35.4
164.9
40
Ø40
Ø40
40
20
2X R20
20
60
2X Ø20
100

50
15
15
70
15
Ø40
VIEW A-A

A-A

EX-182
15
70
15
35.4
Ø40
119.5
30
34.1
39.1
20
100

50
90
70
135
Ø40
30
135
39.1
Ø30
Ø40
34.1
70
40
20
40
104.9
164.9
20

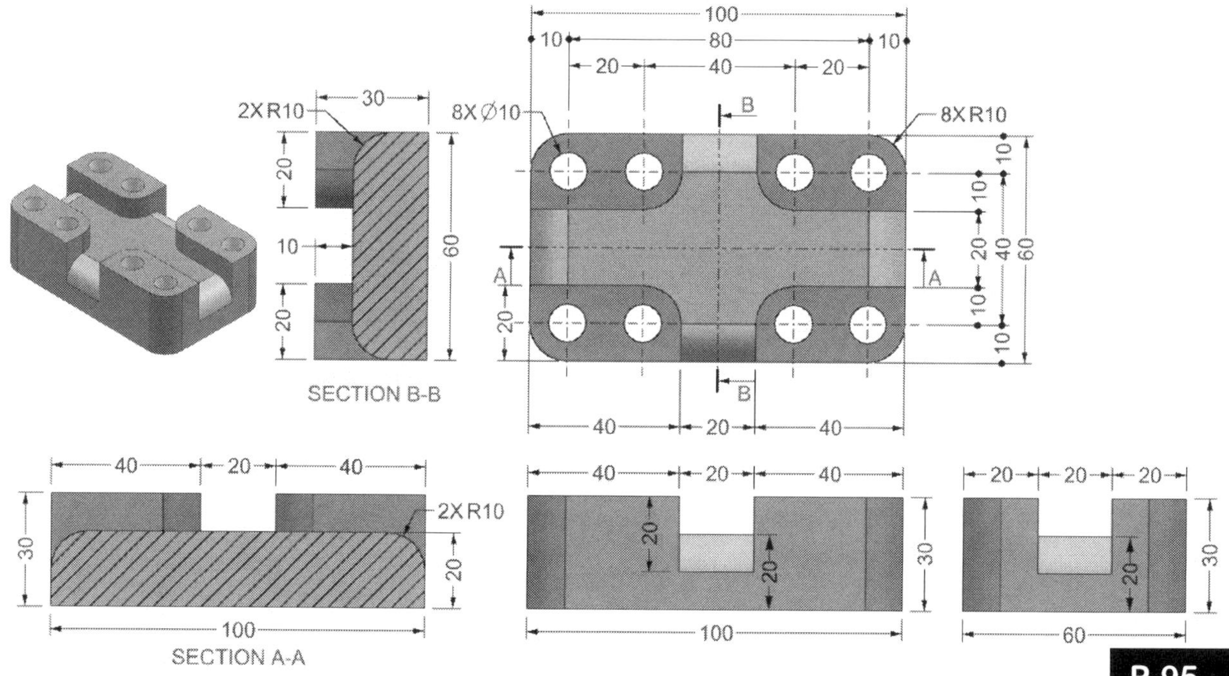

2X R10
30
8X Ø10
100
10
80
10
20
40
20
8X R10
20
B
10
10
60
10
20
40
20
60
10
A
A
20
SECTION B-B
40
20
40
B

40
20
40
2X R10
30
20
100
SECTION A-A

40
20
40
20
20
30
100

20
20
20
20
30
60

P-95

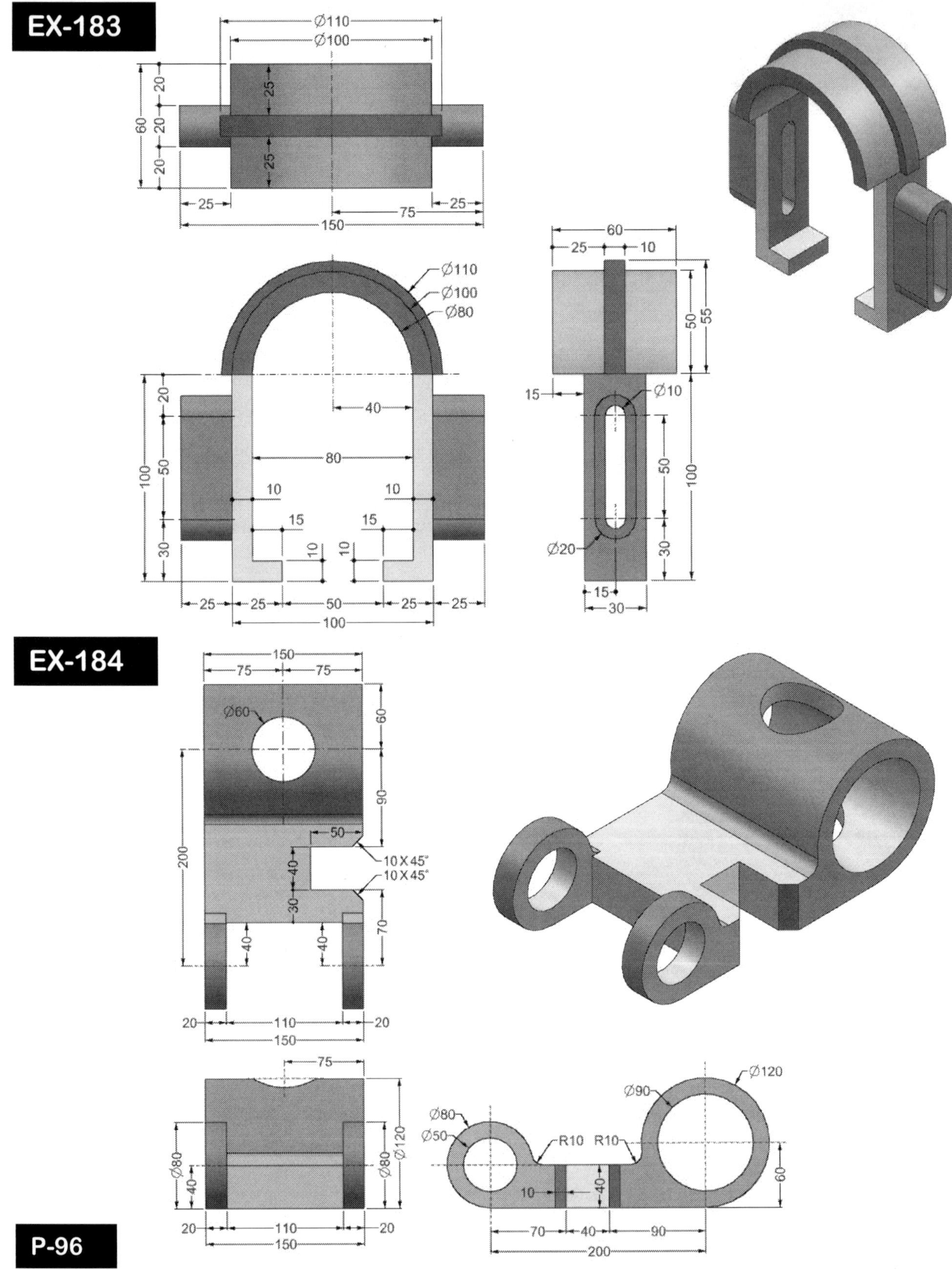
EX-183
Ø110
Ø100
20
20
60
25
20
25
20
25
75
25
150
Ø110
Ø100
Ø80
20
40
50
80
100
10
10
30
15
15
10
10
25
25
50
25
25
100
60
25
10
50
55
15
Ø10
50
100
Ø20
30
15
30
EX-184
150
75
75
Ø60
60
90
200
50
10 X 45°
40
10 X 45°
30
70
40
40
20
110
20
150
75
Ø80
Ø80
Ø120
40
20
110
20
150
Ø120
Ø90
Ø80
Ø50
R10
R10
10
40
60
70
40
90
200
P-96

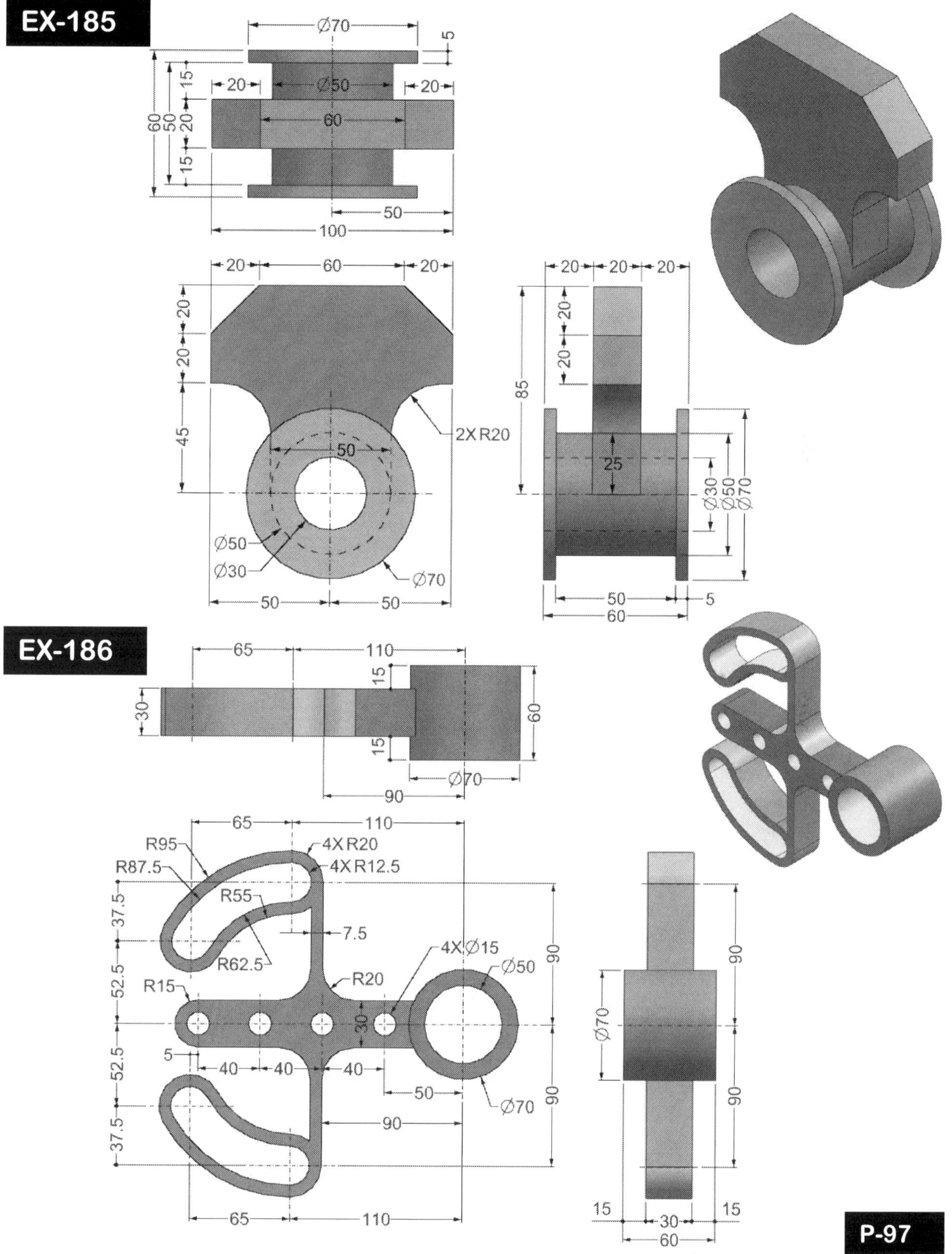

EX-185
EX-186
P-97

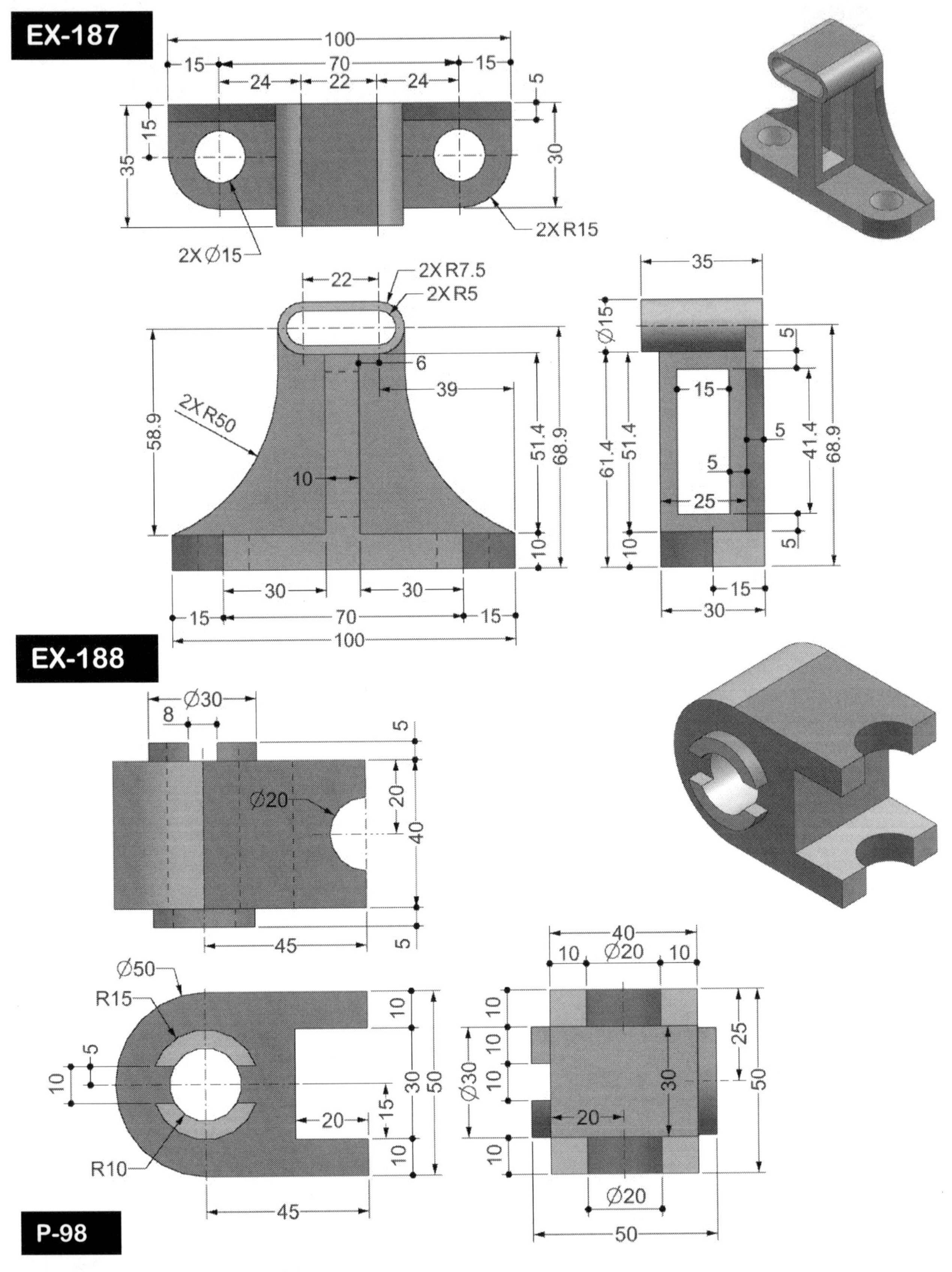

EX-187
100
15
70
15
24
22
24
5
15
35
30
2X R15
2X Ø15
22
2X R7.5
2X R5
6
39
2X R50
58.9
51.4
68.9
10
30
30
15
70
15
100
35
Ø15
5
15
5
61.4
51.4
5
41.4
68.9
25
10
5
15
30
EX-188
Ø30
8
5
Ø20
20
40
45
5
Ø50
R15
10
5
10
30
50
15
20
R10
45
40
10
Ø20
10
10
Ø30
10
10
25
30
50
20
Ø20
50
P-98

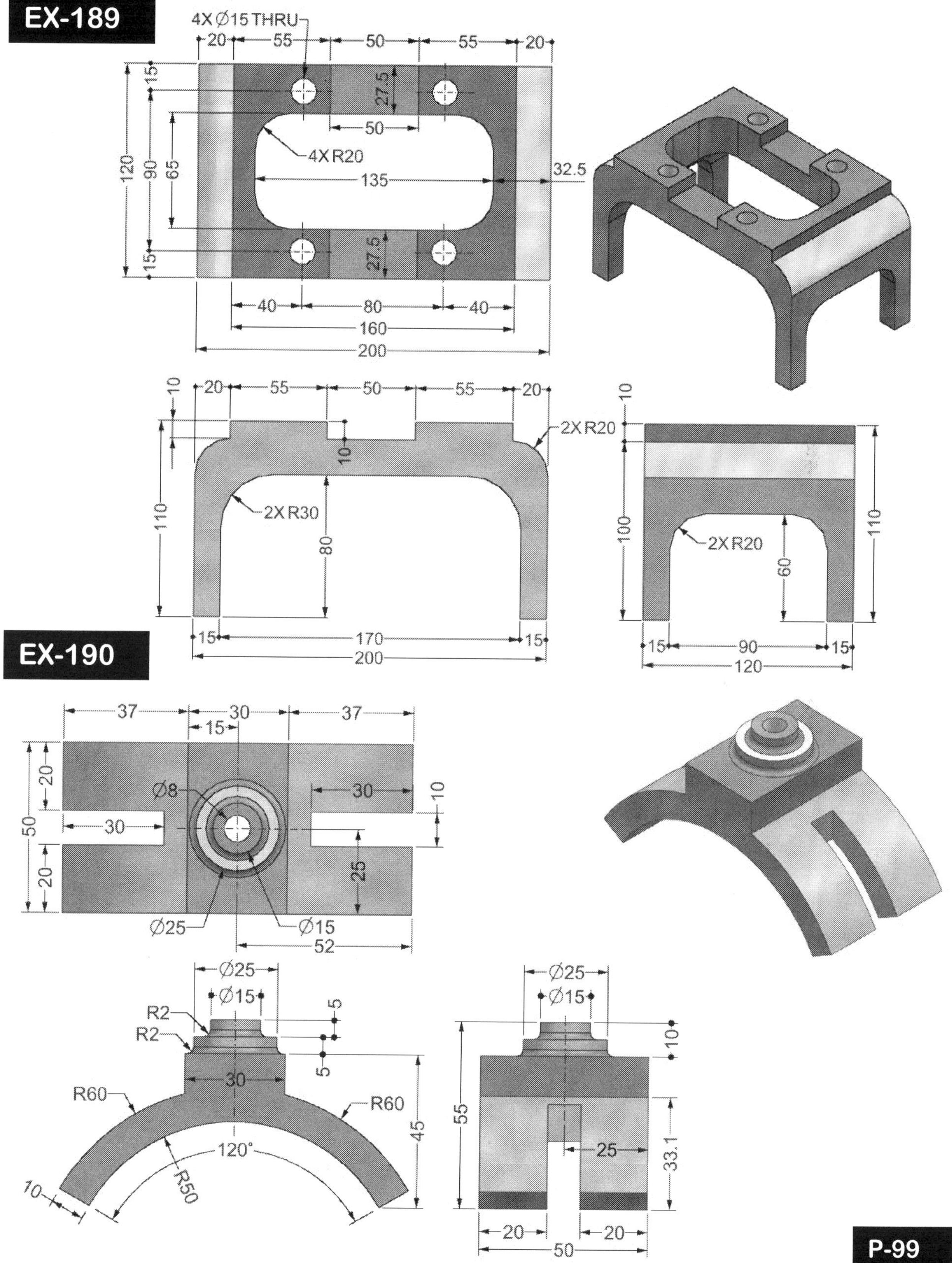

EX-189
4X Ø15 THRU
20
55
50
55
20
15
27.5
50
4X R20
120
90
65
135
32.5
15
27.5
40
80
40
160
200
10
20
55
50
55
20
10
2X R20
110
2X R30
80
15
170
15
200
10
100
110
2X R20
60
15
90
15
120
EX-190
37
30
37
15
Ø8
30
30
10
50
20
20
30
25
Ø25
Ø15
52
Ø25
Ø15
R2
5
R2
5
R60
30
R60
R50
R60
R60
45
120°
10
R50
Ø25
Ø15
10
55
33.1
25
20
20
50
P-99

EX-191 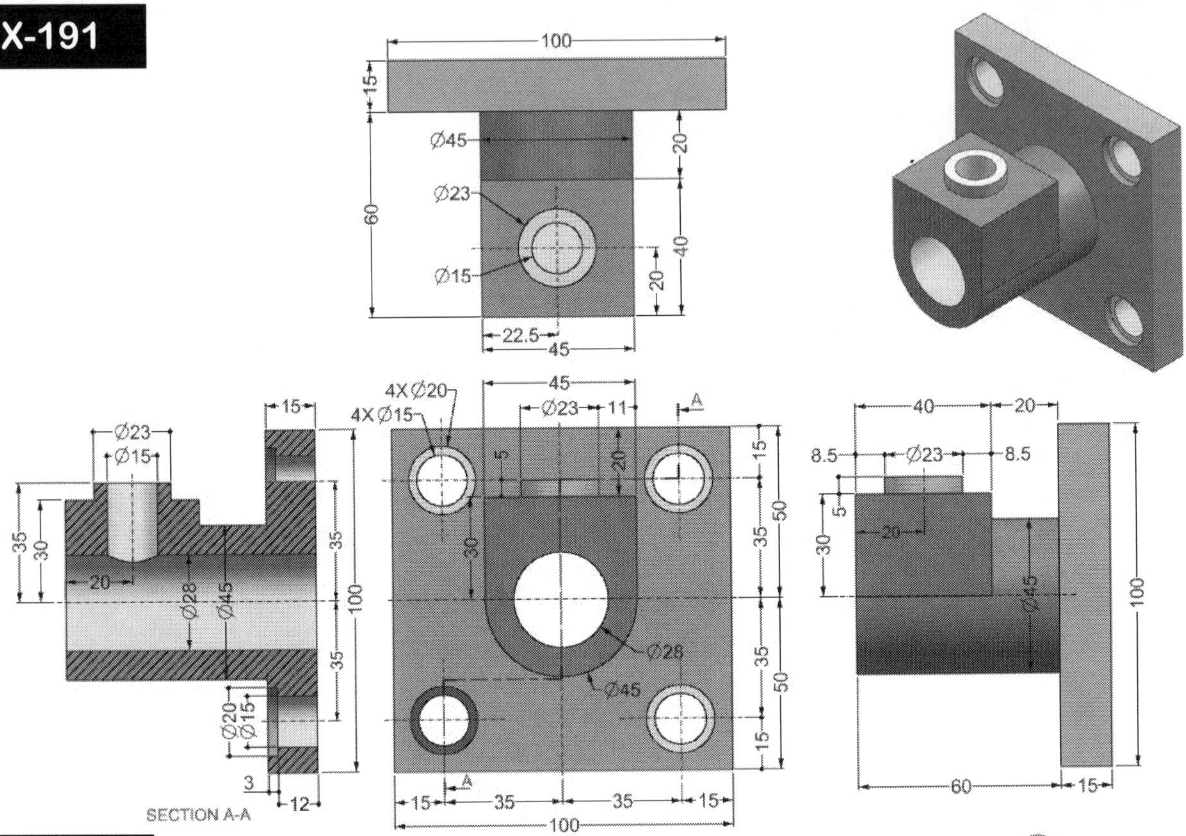
100
15
20
Ø45
40
Ø23
60
Ø15
20
22.5
45
4X Ø20
4X Ø15
45
Ø23
11
A
15
Ø23
Ø15
5
20
35
35
50
50
35
30
100
Ø28
35
35
Ø45
Ø20
Ø15
20
Ø28
Ø45
3
SECTION A-A
12
A
15
35
35
15
100
40
20
8.5
Ø23
8.5
5
30
20
Ø45
100
60
15
EX-192 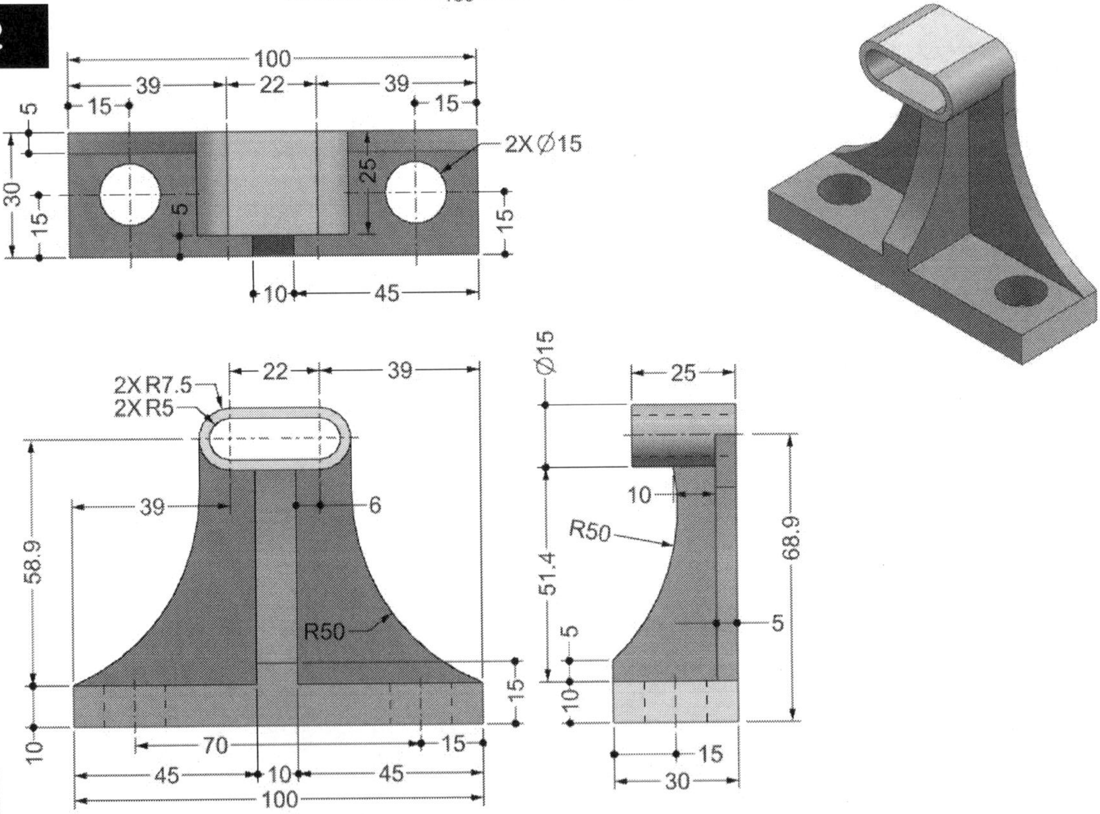
100
39
22
39
5
15
15
25
2X Ø15
30
15
5
15
10
45
Ø15
25
2X R7.5
2X R5
22
39
39
6
10
58.9
51.4
R50
68.9
R50
5
5
15
15
10
70
15
10
45
30
45
10
45
100
P-100

EX-193

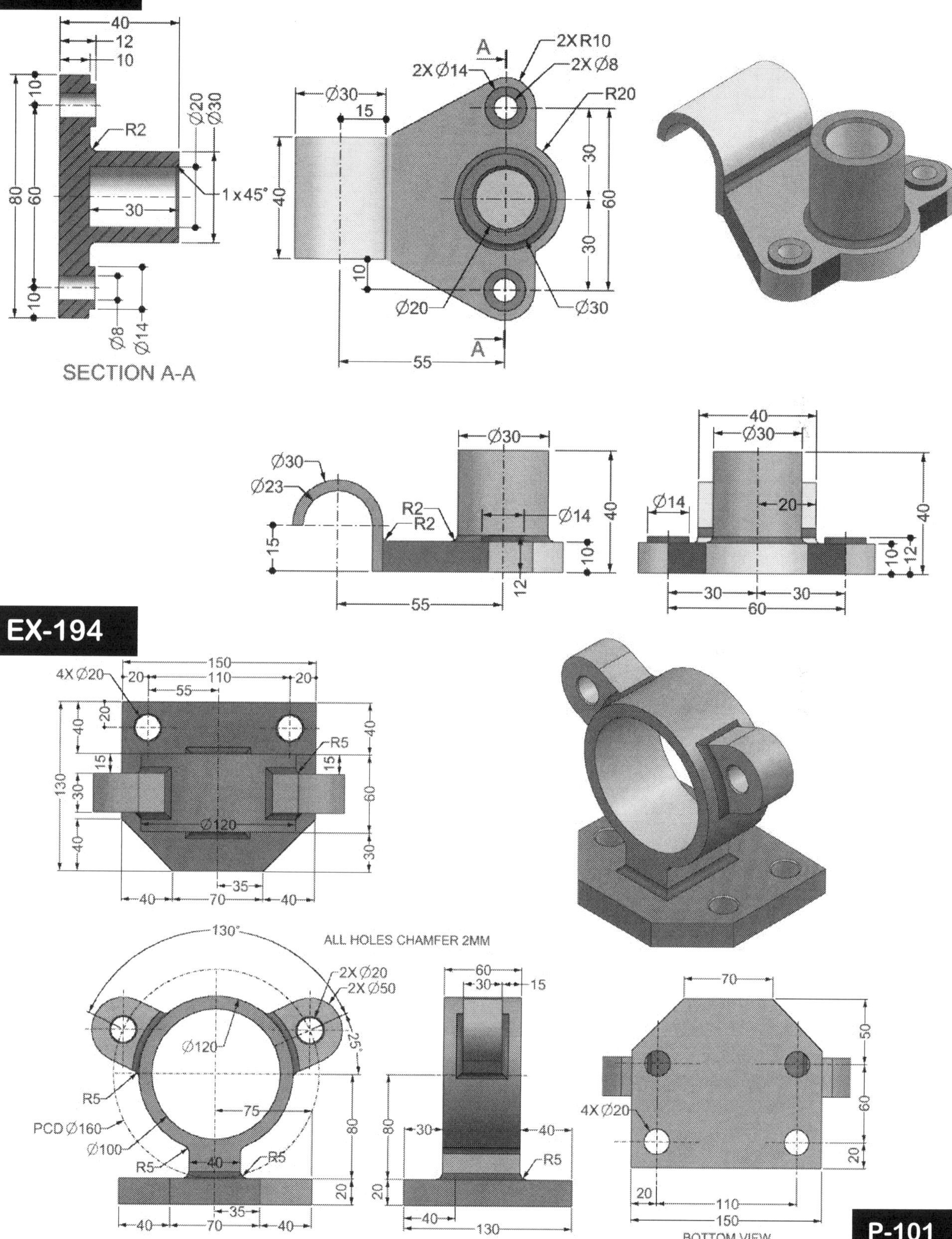
40
12
10
10
80
60
R2
Ø20
Ø30
1 x 45°
30
10
Ø8
Ø14
SECTION A-A

Ø30
15
40
10
55
2X Ø14
A
2X R10
2X Ø8
R20
30
60
30
Ø20
Ø30
A

Ø30
Ø23
R2
R2
15
55
12
Ø14
40
10

40
Ø30
Ø14
20
30
30
60
40
10
12

EX-194

4X Ø20
150
20
110
20
55
20
40
40
130
15
30
R5
15
60
Ø120
40
30
35
40
70
40

ALL HOLES CHAMFER 2MM

130°
2X Ø20
2X Ø50
25°
Ø120
R5
75
PCD Ø160
Ø100
R5
R5
40
20
40
70
35
40

60
30
15
80
30
40
80
20
R5
20
40
130

70
50
60
4X Ø20
20
20
110
150
BOTTOM VIEW

P-101

EX-195

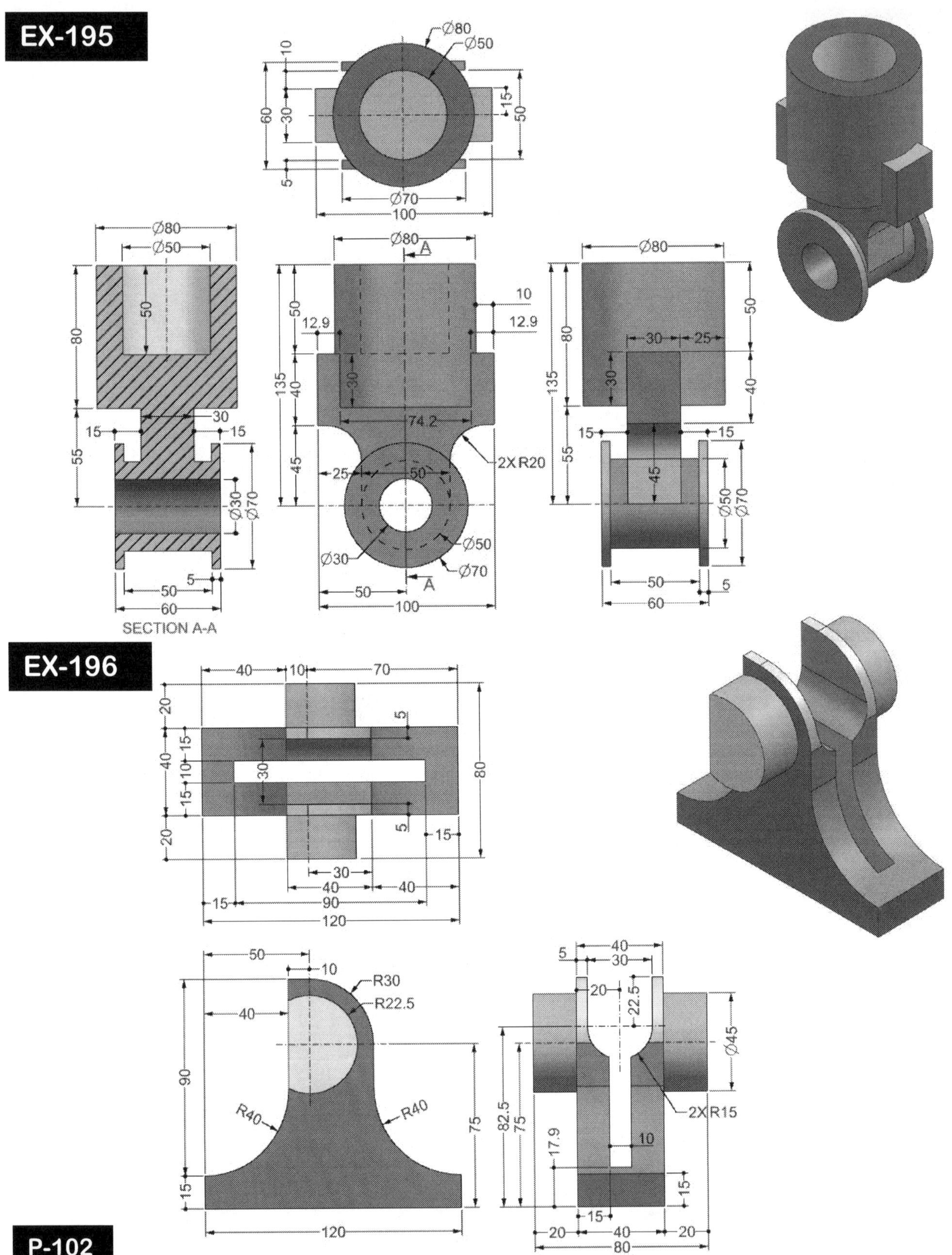
EX-195
Ø80
Ø50
10
60
30
5
50
15
Ø70
100
Ø80
Ø50
50
80
30
55
15
15
Ø30
Ø70
50
5
60
SECTION A-A
Ø80
A
135
50
12.9
40
45
30
74.2
25
50
2X R20
Ø30
Ø50
Ø70
50
100
A
10
12.9
Ø80
80
30
25
135
50
30
40
55
45
15
15
Ø50
Ø70
50
5
60
EX-196
40
10
70
20
5
15
40
10
30
15
80
5
15
20
30
40
40
15
90
120
50
10
R30
R22.5
40
90
R40
R40
75
120
5
40
30
20
22.5
82.5
75
Ø45
17.9
10
2X R15
15
15
20
40
20
80
P-102

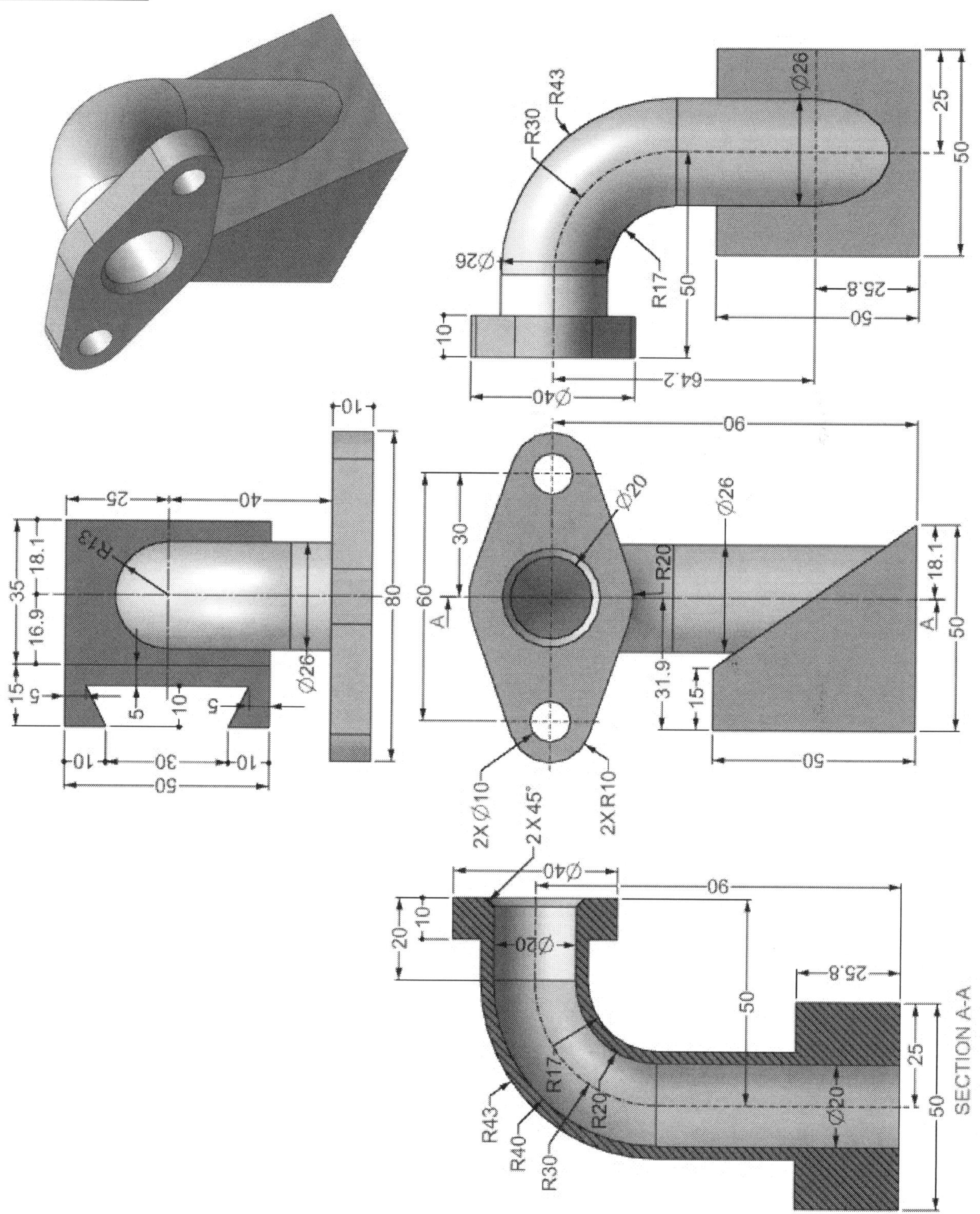
Ø26
R43
R30
Ø26
R17
50
10
Ø40
64.2
25
50
25.8
50
Ø26
10
25
40
R13
35
18.1
16.9
15
5
5
10
5
Ø26
80
10
30
10
50
60
30
A
Ø20
Ø26
R20
A
18.1
50
31.9
15
50
90
2X Ø10
2 X 45°
2XR10
Ø40
20
10
Ø20
90
25.8
R17
R43
R40
R30
R20
50
Ø20
25
50
SECTION A-A

EX-198

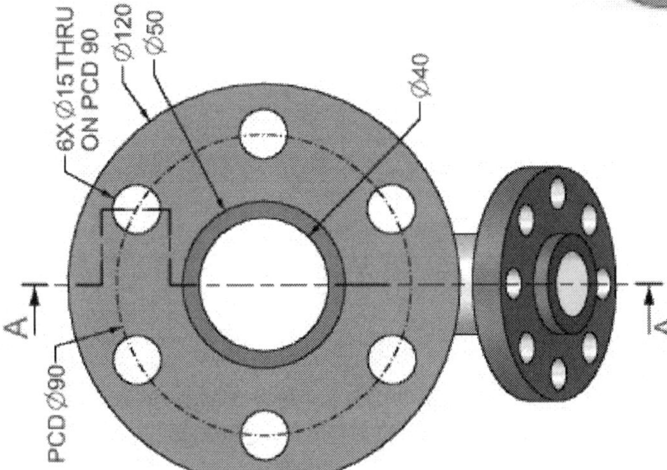

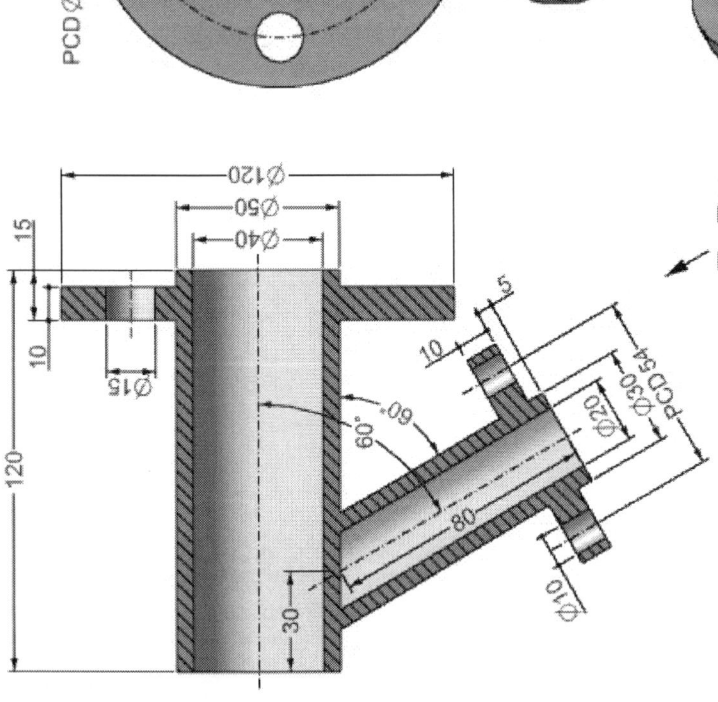

6X Ø15 THRU
ON PCD 90

Ø120
Ø50
Ø40

PCD Ø90

A
A

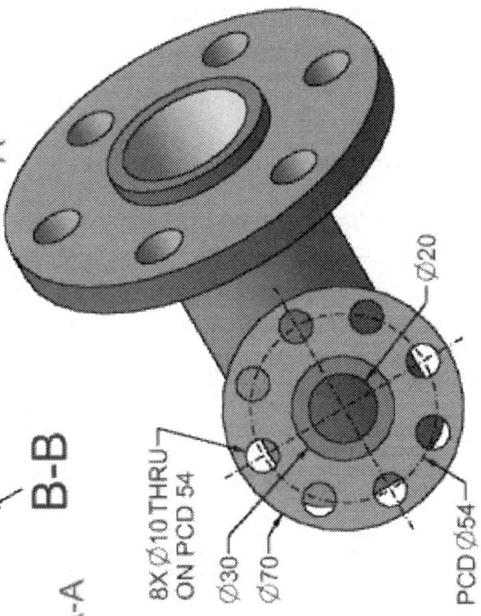

VIEW B-B

8X Ø10 THRU
ON PCD 54

Ø20
Ø30
Ø70
PCD Ø54

B-B

Ø120
Ø50
Ø40
15
10
Ø15
120
30
60°
60°
80
5
10
Ø20
Ø30
PCD 54
Ø10

SECTION A-A

P-104

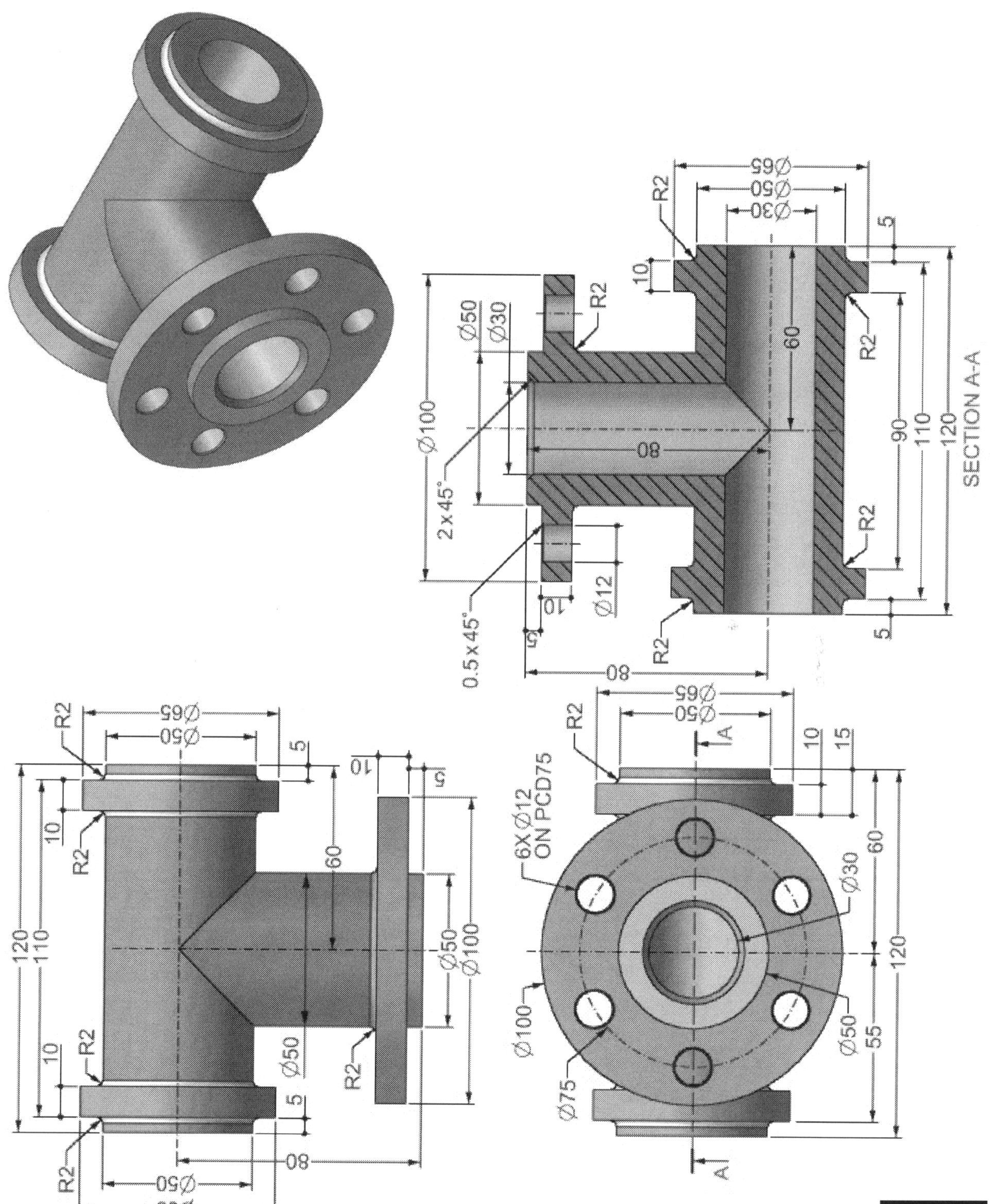

SECTION A-A
R2
Ø65
Ø50
Ø30
5
10
60
90
110
120
R2
R2
Ø50
Ø30
R2
Ø100
2 x 45°
80
0.5 x 45°
10
Ø12
R2
5
80
R2
Ø65
Ø50
5
10
R2
60
Ø50
Ø100
R2
120
110
10
R2
5
Ø50
R2
80
Ø50
Ø65
R2
Ø65
Ø50
A
10
15
6X Ø12
ON PCD75
Ø30
60
Ø100
Ø50
55
Ø75
120
A

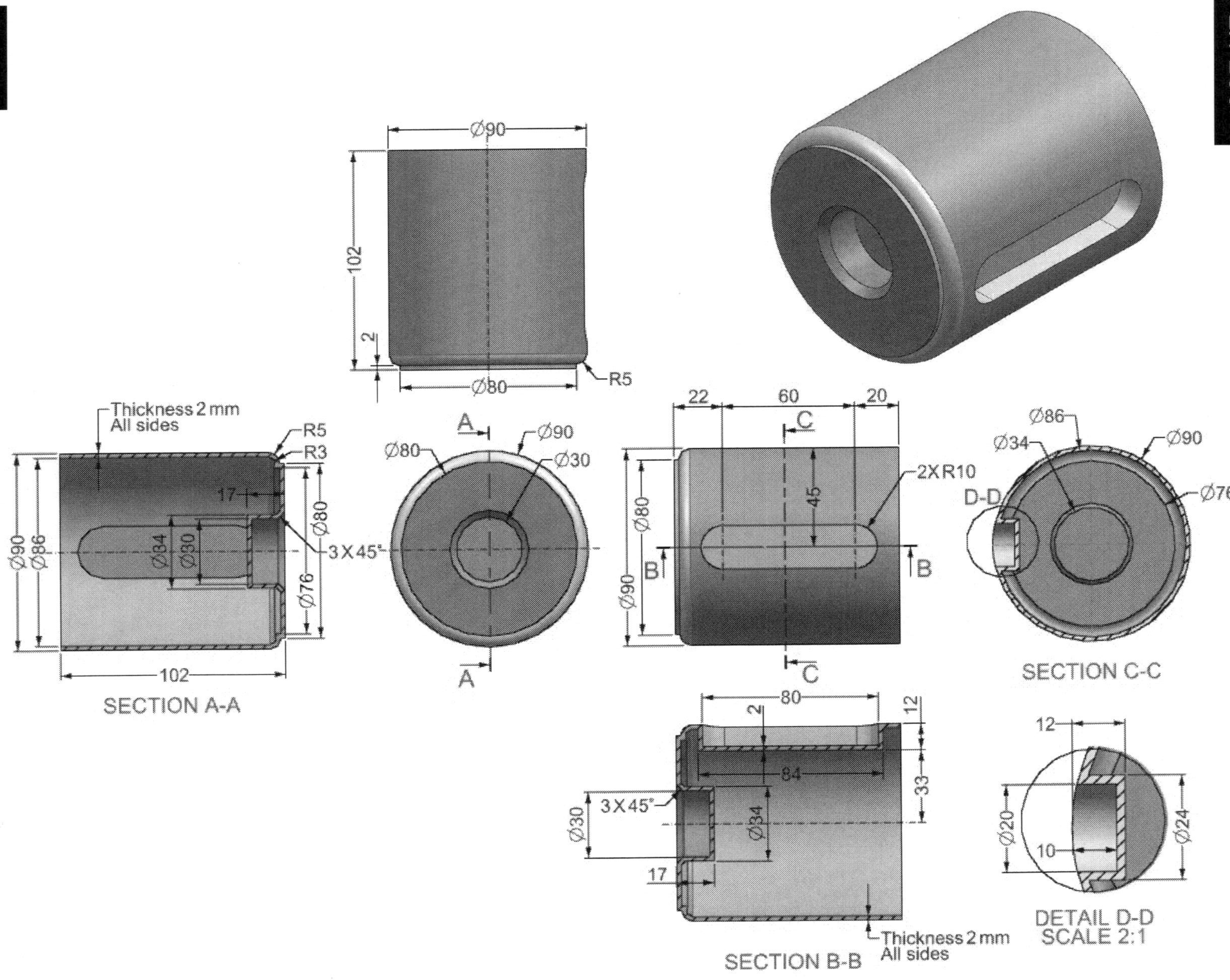

P-106
Ø90
102
2
Ø80
R5
Thickness 2 mm
All sides
R5
R3
17
Ø80
Ø90
Ø86
Ø76
3 X 45°
Ø34
Ø30
102
SECTION A-A
Ø80
A
Ø90
Ø30
A
22
60
20
C
45
Ø80
Ø90
2X R10
B
B
C
SECTION B-B
80
2
12
84
33
Ø30
3 X 45°
Ø34
17
Thickness 2 mm
All sides
Ø86
Ø34
D-D
Ø90
Ø76
SECTION C-C
12
Ø20
10
Ø24
DETAIL D-D
SCALE 2:1

Other useful books by CADIN360

1. 150 CAD Exercises

2. AutoCAD Exercises

3. CAD Exercises

4. 50+ SolidWorks Exercises

5. SolidWorks 200 Exercises

6. Autodesk Inventor Exercises

7. Catia Exercises

8. Siemens NX Exercises

Made in the USA
Monee, IL
07 May 2026